Malaysia International NDT Conference and Exhibition (MINDTCE'18)

IOP Conference Series: Materials Science and Engineering
Volume 554

Subang Jaya, Malaysia
12 – 13 November 2018

ISBN: 978-1-5108-8814-2
ISSN: 1757-8981

Printed from e-media with permission by:

Curran Associates, Inc.
57 Morehouse Lane
Red Hook, NY 12571

Some format issues inherent in the e-media version may also appear in this print version.

Printed by Curran Associates, Inc. (2019)

For permission requests, please contact the Institute of Physics at the address below.

Institute of Physics
Dirac House, Temple Back
Bristol BS1 6BE UK

Phone: 44 1 17 929 7481
Fax: 44 1 17 920 0979

techtracking@iop.org

Additional copies of this publication are available from:

Curran Associates, Inc.
57 Morehouse Lane
Red Hook, NY 12571 USA
Phone: 845-758-0400
Fax: 845-758-2633
Email: curran@proceedings.com
Web: www.proceedings.com

Table of contents

Volume 554

Malaysia International NDT Conference and Exhibition 2018

12–13 November 2018, Sunway Pyramid Convention Centre, Malaysia

Accepted papers received: 16 May 2019

Preface

Papers

IOP Conf. Series: Materials Science and Engineering **554** (2019) 011001 doi:10.1088/1757-899X/554/1/011001

Preface

Organizing conference on a very specialize field in this country is really challenging effort. This fact is also true when organizing the 'Malaysia International NDT Conference and Exhibition 2018 (MINDTCE'18) on 12 and 13 November 2018. Persuading national and international members of NDT experts to participate and to share their knowledge and experience in such a conference is considered as the biggest challenge that has to be overcome by members of organizing committee. For this reason, when the event was successfully held with an excellent response from the presenters, exhibitors and normal participants, every individual involved on the organization of this event were so relieve and delighted. Thanks and congratulation to all especially members of Steering Committee who has spent days and nights making efforts to ensure that the event executed in the best possible manner.

Members of Organizing Committees thanks all presenters who has agreed to have their papers that contain extremely valuable information and data that took years for them to collect and produce. In order to enhance their quality, these papers were edited by a number of national and international experts. The organizers thanks the publisher, IOP Publishing who values these papers qualified to be published in Scopus indexed journals of Material Science and Engineering. We hope this publication add to one more source of rarely published knowledge in the field of NDT that are needed by researchers and practitioners throughout the world in their strive to enhance the role of NDT in creating safer and cleaner work environment for all mankind.

IOP Conf. Series: Materials Science and Engineering **554** (2019) 011002 doi:10.1088/1757-899X/554/1/011002

Peer review statement

All papers published in this volume of *IOP Conference Series: Materials Science and Engineering* have been peer reviewed through processes administered by the proceedings Editors. Reviews were conducted by expert referees to the professional and scientific standards expected of a proceedings journal published by IOP Publishing.

IOP Conf. Series: Materials Science and Engineering **554** (2019) 012001 doi:10.1088/1757-899X/554/1/012001

A Feature Extraction Technique Based on Factor Analysis for Pulsed Eddy Current Defects Categorization

Nurul A'in Ahmad Latif [1,3], Ilham Mukriz Zainal Abidin[1], Noorhazleena Azaman[1], Nordin Jamaludin[2] and Ainul Akmar Mokhtar[3]

[1]Leading Edge Non Destructive Testing Group, Industrial Technology Division, Malaysian Nuclear Agency, 43000 Bangi Selangor, Malaysia
[2]Department of Mechanical and Materials Engineering, Faculty of Engineering and Built Environment, Universiti Kebangsaan Malaysia,43600 Bangi, Selangor, Malaysia
[3]Mechanical Engineering Department, Universiti Teknologi Petronas, 32610 Bandar Seri Iskandar, Perak, Malaysia

Corresponding author, email: nurul_ain@nm.gov.my

Abstract. Pulsed eddy current (PEC) is an advance Non-destructive testing that is widely used in multiple industries for surface and subsurface defect detection. Normally, the conventional features extractions are conducted based on time to peak and peak value response. However, signals extraction could be a confusing for a thin specimen because times to peak for both surface and subsurface defects are occurred at almost similar time. This paper introduced the application of confirmatory Factor Analysis for defect categorizing for stainless steel thin plate. Through this statistical method, two categories of defects namely surface and subsurface defects are clearly to be distinguished between one to another.

1. Introduction

Pulsed eddy current (PEC) is an emerging technique and operates based on electromagnetic principle and commercially applied in multi discipline fields including aeronautical, manufacturing and petroleum industries. PEC is categorized as an advance Non-destructive testing (NDT) develops for defect quantification including surface and subsurface crack measurement, depth estimation and crack construction. In contrast to conventional eddy current that excites by single or harmonic excitation at one frequency value, PEC is excites by square waveform. The advantages of using continuous spectrums are it allowed a deeper penetration and it is thoughtful to bring information about anomalies that are deeply buried inside the specimen and higher robustness of anti-interference [1].

In PEC, the peak value and peak time are the main features used for signals interpretation. Peak values are correlated to the size of the defect whereas the latter are representing the depth of the defect. Tian and Sophian [2] proposed a new feature for defect classification termed as rising time. Rising time feature is used to identify the types of defects and overcome the lift-off issue. Since PEC system are developed from wide-band electronics device thus it may impair and prone to signal-to-noise. It may require a complex data analysis for signal interpretation compare to eddy current conventional system. Recently, it has gained an attention and requirements from industry to develop an accurate signals analysis for ensuring the integrity of tested structure.

In this research, the specimen used is stainless steel 304 with nominal thickness 0.051 mm. The specimen selection is based on the material and parameters that representing the fuel cladding of TRIGA

MARK II Nuclear Reactor. The signals interpretation and characterization for very thin specimen may have a bit confusing due to the difficulty to distinguish between the surface and subsurface defects. The possibilities for time to peak value for both defects to be occurred at the same time are relatively very high.

Predominantly, statistical analysis generates important information from research conducted including patterns, characterization and demographics. In 2002, Sophian et. al [3] introduced the application of principal components analysis (PCA) for defect classification and quantification based on PEC response acquired. In this research, a new signals analysis is introduced based on statistical method, factor analysis. Generally, PCA and FA are looks like very similar and familiar in many ways. However, they have different in the term of its fundamental that will bring a huge effect on its applications. The function of PCA is to perform data reduction and the purposes are to create one or more index from one large population measured. These created index variables are called components. Figure 1 shows the fundamental operation processes of PCA in data reducing.

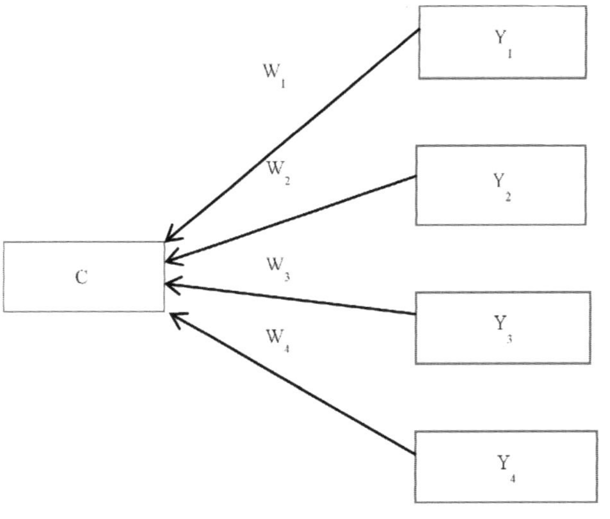

Figure 1. Fundamental operation of PCA in data reducing.

Based on this figure, PCA is combining 4 measured variables (*Y*) into single component, *C*. It is clearly shown that the *Y* variables contribute to the components variable, *C* and this model can be simplified using Eq. 1.

$$C = W_1Y_1 + W_2Y_2 + W_3Y_3 + W_4Y_4 \quad (1)$$

where W_n (n = 1, 2, 3 and 4) = weight or factor score coefficient

In contrast to PCA, factor analysis measures a latent variable that is cannot to be measured from one single variable. The measurement for FA is illustrated as shown in Fig. 2. Theoretically, factor analysis is a multivariate analysis method that used to identify the underlying factors that are correlated between covariation among the independence variables. The objective of its implementation is to reduce the number of variables used to explain any linkages or relationship establish among the variables. From the figure it describes that the factor, F is causing the responses on the four measured variables, Y. This model literally can be interpreted as a set of regression and can be calculated using Eq. 2.

IOP Conf. Series: Materials Science and Engineering **554** (2019) 012001 doi:10.1088/1757-899X/554/1/012001

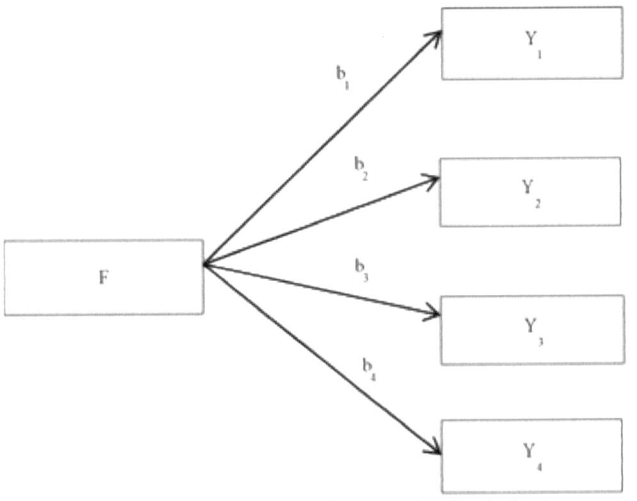

Figure 2. Fundamental operation of Factor Analysis in data reducing.

$$Y_1 = b_1 F$$
$$Y_2 = b_2 F$$
$$Y_3 = b_3 F$$
$$Y_4 = b_4 F$$

where b_n (n=1, 2, 3 and 4) = factor loading

Factor analysis has been reported was used in multi disciplines research including construction [4], groundwater quality inspection [5] and financial analysis [6]. However, to date is never been reported about its application in any of non-destructive testing and inspections. This research used a confirmatory factor analysis approach to determine the relationships between the variables selected which are time to peak and peak value. Confirmatory approach will give a confirmation whether the selected measurement fitted the scope of the research.

The rest of the paper is organized as follows. A section 2 discusses about the factor analysis approaches. Section 3 introduces the experiment works conducted and results are briefly discuss in Section 4 and finally in Section 5 is the conclusion.

2. Application of factor analysis for experimental data

In our study, we investigate the application of factor analysis in categorizing defects into surface and subsurface based on the variables and compare this statistical results with the normal signal analysis outcome.

Factor Analysis is a commonly operates based on variable reduction technique. This multivariate statistical technique is used for three primary reasons including to reduce the number of variables, from large to small and also to establish underlying dimensions between measured variables and constructs. In addition, it is also very useful for providing construct validity evidence.

We apply a confirmatory Factor Analysis based signal processing using PEC developed system to extract the features for defect detection and classification. As described in Section 1, FA is measuring the latent variable that cannot be calculated from one single variable by itself. In this experiment, the two identified factors to be measured are surface and subsurface defects. For the defect quantification, the common signal features, time to peak and peak value are analysing in order to find the correlation between these variables and to establish the underlying factors that linked to each other. By applying the FA for a signals analysis, the defects that are identified as are factors are more easily to be distinguished based on the correlation established between selected variables.

IOP Conf. Series: Materials Science and Engineering **554** (2019) 012001 doi:10.1088/1757-899X/554/1/012001

According to Cunningham [7], the factor analysis method is utilized to determine the nature of the construct influencing a set of responses and to achieve the validity of data purposes. Thus this research followed Ramani [8] who indicated that it is the simplest method to explore the constructs. Initially, there are four guidelines to be performed in ensuring that the data are well cooperate and suitable for factor analysis statistical method. The four steps to be conducted are including The Kaiser-Meyer-Olkin (KMO) test, Bartlett's test, variance measurement and factor rotation.

The Kaiser-Meyer-Olkin (KMO) is the test conducted to measure how suited the data to be performed using Factor Analysis and it measures the sampling adequacy for each variable. This statistical test is measuring the proportion of variance among variables that might be common variance. The accepted KMO values are equal or more than 0.50. If the KMO value acquired is less than 0.5, it means that the data is not suitable for this statistical analysis. KMO values between 0.5 and 0.7 are considered as mediocre, 0.7 and 0.8 as good, values between 0.8 and 0.9 are great and values above 0.9 as superb [9].

The second statistical test performed is Bartlett's test. The Bartlett's test of sphericity was used to test the significance of correlations among all factors, with 0.05 cut off point employed in determining the significance level.

Factor Rotation (loading factor) - The process of adjusting the factor axes in order to get a simpler and more significant factor solution. The rotated values should be more than 0.40. If no factor rotation appeared, this means the measurements are already significantly simple. The item (s) need to be deleted as measurement in variable if loading value is less than rotated threshold. The data can only be recognized to perform factor analysis if all the results from these four guidelines are fulfilled.

The purpose of the factor analysis rule above is to achieve data reduction or retain the nature and character of the original items, and to delete those items which had lower factor loadings and cross loading (10).

3. Case Study

3.1 Finite element model

The signals analysis conducted for approach, Factor Analysis and conventional signal analysis are performed using simulation works' results. This finite element method work is simulates using a dedicated software, COMSOL Multyphysics in 3-dimension geometry. The main reason of selecting a simulation approach instead of experiment work is to establish the ideal environment workspace thus it is expected producing an accurate result (discrepancies). Hence the results are free from any insignificants errors contributed by electronics noises, external magnetic field disturbance and human errors in which it may leads into a misleading interpretation. In this simulation works the specimen used is 0.50 mm thickness stainless steel 304 plate fabricated with eight different depths and the opening size of each defect is 2.0mm. The defects depth, *d* are 0.15mm, 0.20 mm, 0.25mm, 0.30mm, 0.35mm,0.40mm,0.45mm and 0.50mm as illustrated in Fig 3. The time step of the pulse waveform is 10 ms with 5 ms excitation pulse width and pulse repetition frequency is 100 Hz.

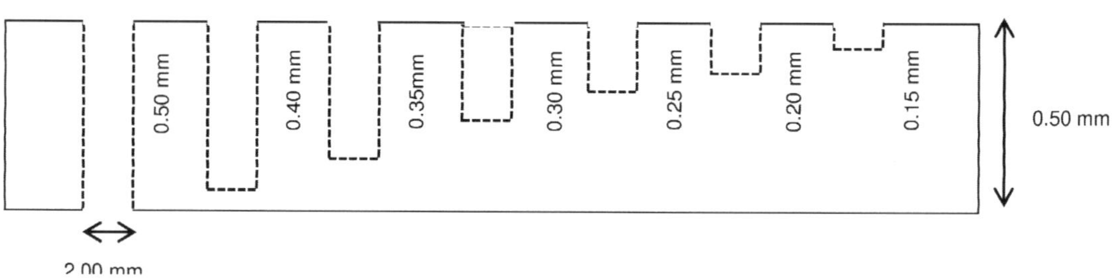

Figure 3. Illustration of stainless steel 0.50 mm plate thickness fabricated with seven different defects depth.

3.2 EC probe
PEC probe is generally robust, small and provides a high sensitivity in detection. In this simulation work, the probe has dimension of 16 mm outer diameter, 10.5 mm inner diameter, 30 mm height and made up from 300 turns of copper wire. Figure 4 illustrated the schematic diagram of PEC probe.

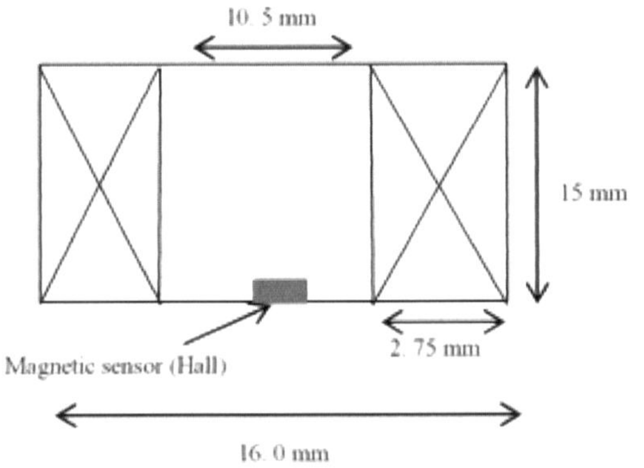

Figure 4. Schematic diagram of PEC probe.

For the surface detection, the probe will be positioned above the defect opening with a lift-off of 2 mm. Meanwhile for subsurface defect detection, the probe is positioned on the other side of the plate. The orientation for these finite element method works are shown in Figure 5a and 5b. All the data acquired from these simulation works will be analysed using statistical factor analysis method and conventional signals analysis approach for comparison purposes.

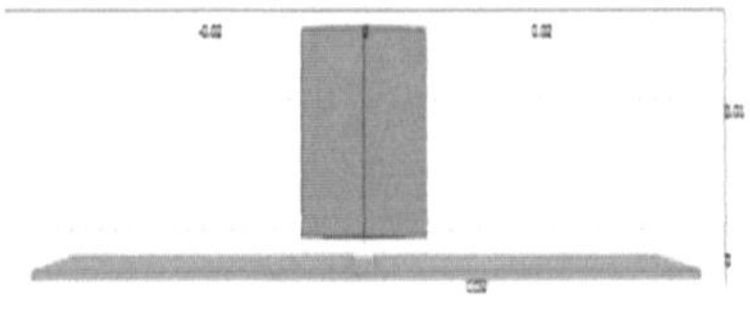

Figure 5a. Finite element method work on surface defect.

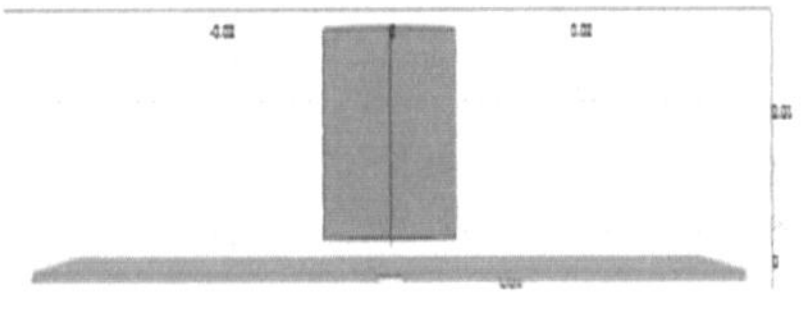

Figure 5b. Finite element method work on surface defect.

4. Results

4.1 *Simulation work*
In this paper, we are going to categorize the defects based on their location in the specimen conducted through two different approaches, finite elements work and factor analysis.
Figure 6a and 6b show the results acquired from simulation works conducted on surface and subsurface defects. From the results it shows that the deepest defect having the lowest net magnetic field and the scenario is vice versa for the shortest depth. Based on the results, the magnetic field value for each defect depths show a similarity regardless their location either on surface or subsurface. For example, Figure 7 shows that the 0.20 mm defect depth for both surface and subsurface generates magnetic field relatively same around 0.018 T. From Fig. 6a and 6b, the peak value for surface and subsurface at 0.20 mm depth are 0.01843 T and 0.01850T. Meanwhile the time to peak are 0.1742 ms and 0.1751 ms. Peak value for both types of defects are relatively same for one particular depth and the time to peak is happened at almost same time. These kind of signals condition will lead into confusing especially for interpretation and defect categorization based on their location. To overcome this problem, the peak and time to peak values for both types of defects are extracted and analysis using factor analysis.

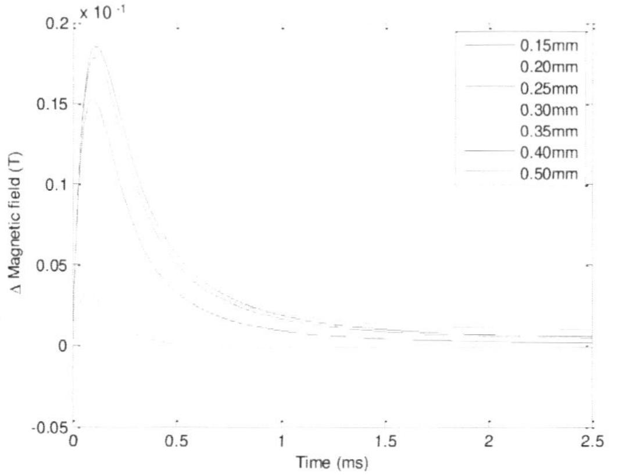

Figure 6a Transient net magnetic field acquired from surface defects.

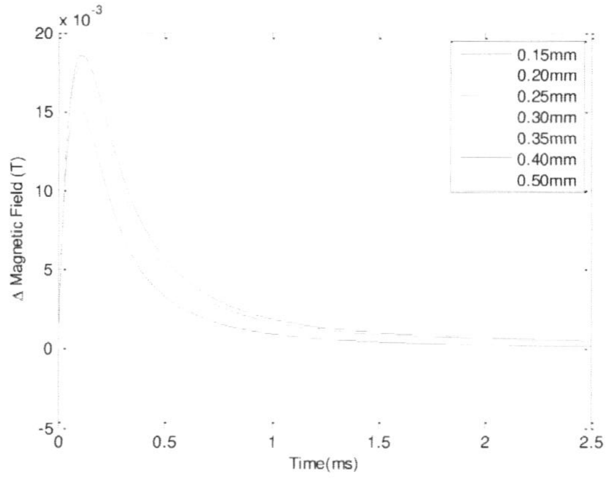

Figure 6b Transient net magnetic field acquired from subsurface defects.

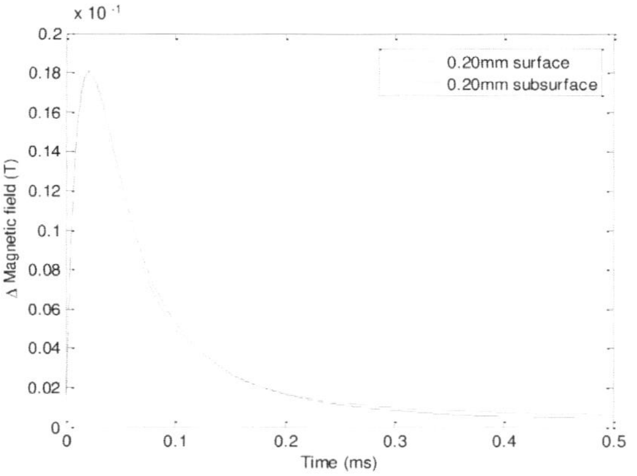

Figure 7. Transient net magnetic field acquired from 0.20 mm surface and subsurface defect.

4.2 *Factor analysis*

FA is conducted to categorize the defects based on peak value and time to peak values. For each types of defect, 8 measurements are taken giving it a total of 128 measurements. According to Hatcher [11] for factor analysis approaches the number of sample size should be of at least 100 or 5 times the number of variables.

IOP Conf. Series: Materials Science and Engineering **554** (2019) 012001 doi:10.1088/1757-899X/554/1/012001

Table 1 show the factor analysis results performed using the simulation data collected. In accordance to the requirements, it is mandatory to identified either the sampling is adequate for the analysis to be conducted. Table 1 shows the KMO, Bartlett's Test and variance for the whole set of data.

Table 1. Kaiser-Meyer-Olkin, Bartlett's Test and variance measurements acquired from finite elements work conducted on surface and subsurface defects.

<table>
<tr><th rowspan="2">Factor</th><th colspan="2">Component</th></tr>
<tr><th>1</th><th>2</th></tr>
<tr><td>pv1</td><td></td><td>0.945</td></tr>
<tr><td>ttp1</td><td></td><td>0.945</td></tr>
<tr><td>pv2</td><td>0.979</td><td></td></tr>
<tr><td>ttp2</td><td>0.979</td><td></td></tr>
<tr><td>% of variance</td><td>49.573</td><td>43.009</td></tr>
<tr><td>Kaiser-Meyer-Olkin (KMO)</td><td colspan="2">0.50</td></tr>
<tr><td></td><td colspan="2">Bartlett's Test of Sphericity</td></tr>
<tr><td>Approx. Chi Square</td><td colspan="2">148.251</td></tr>
<tr><td>df</td><td colspan="2">6</td></tr>
<tr><td>Significant</td><td colspan="2">0.0000</td></tr>
</table>

Component 1 is the subsurface defect and it represents time to peak (ttp1) and peak value (pv1). Meanwhile component 2 is representing surface defect having of two variables, time to peak (ttp2) and peak value (pv2).

The KMO is measuring the sample adequacy and it varies between 0 and 1. The value that closer to 1 is better and 0.50 is set as minimum value as described by Ramani [12]. From the table, the KMO (p) value calculated is 0.50 thus the sampling data is accepted to be proceeded using factor analysis. Meanwhile, a Bartlett's Test of Sphericity is the test for null analysis used to determine either the correlation matrix is an identity matrix. The purpose of using Bartlett'd Test of Sphericity is to identify whether the variables are load properly. According to the law, an identity matrix is where the diagonals are all 1 and the off-diagonals are all 0. There are two hypotheses assume for this analysis as stated below:

i. Null Hypothesis, H_0: There is no statistically significant interrelationship between variables and type of defects.

ii. Alternate Hypothesis H_1: There may be a statistically significant interrelationship between variables and type of defects.

Taking a 95% of level of Significance, α=0.05. From Table 1, the p-value (Sig) 0.00< 0.05, thus the Factor Analysis is valid. Since $\rho<\alpha$, therefore we reject the H_0 and accept the H_1 that there is a statistically significant interrelationship between variables and type of defects. Meanwhile the minimum value for variances as mentioned by Ramani is 50%. In Table 1, the total variance value calculated is 92.582 %. Thus the data is valid for factor analysis statistical approach.

Figure 8 shows the scree plot obtained from the simulation conducted on both surface and subsurface defects. From the scree we can determine how many numbers of the factors can be included into the model by looking at the scree plot begin to level off. Additionally, any factors with Eigenvalue less than 1 need have to be excluded from the model. The Eigenvalue used to establish the cut off of the factors involve represent the strength of each factor. Based on this figure, there are two components, component 1 and 2 with Eigenvalue more than 1. It is clearly shows that surface and subsurface defects are the variables that explain at least an average amount of the variance.

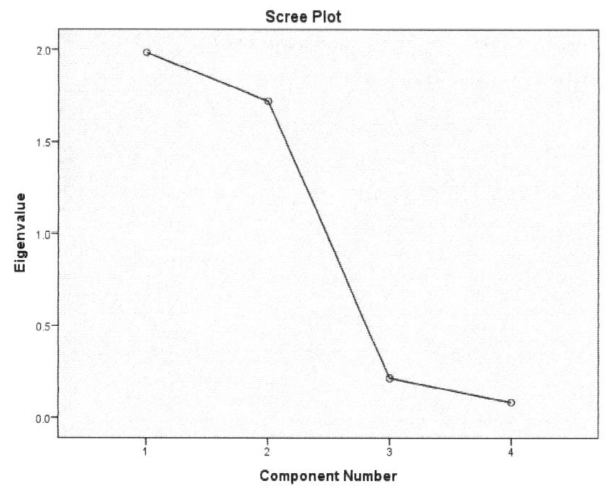

Figure 8. Scree plot obtained from the simulation works.

Meanwhile Figure 9 shows the components plot in rotated space using orthogonal Varimax with Kaiser Normalization and rotation converged in 3 iterations. In Varimax rotation, the values less than 0.40 had been blanked out. In fact, the higher factors loadings indicate that the particular variable is closely associated with the monitored factor. Table 2 shows the rotated component matrix for both components, surface and subsurface defects. Based on the table, times to peak and peak value for subsurface defect are highly associated and can be loaded into same group of factor. Meanwhile, times to peak and peak value for surface defect with factor value, 0.945 are highly related and can be clustered into one factor group.

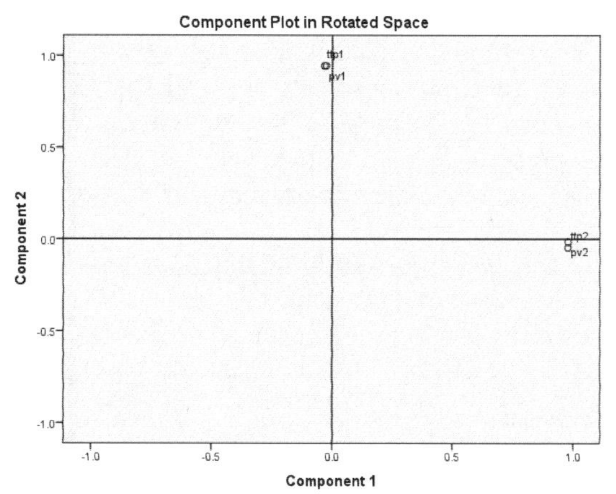

Figure 9. Component plot in rotated space.

Table 2. Rotated component matrix.

Factor	Component	
	1	1
pv2	0.979	
ttp2	0.978	
pv1		0.945
pv2		0.945

5. Conclusion

A new feature extraction of PEC signals using Factory Analysis analytical method has been developed and investigated. The KMO measurement and Bartlett's test show that the data are significantly related.

This statistical method focusing on the defects categorization based on the eigenvalue on the Scree plot and component plot in rotated space. These values indicate the differences of the defects classes which in this experiment both are represent surface and subsurface defects. In future, it is suggested to conduct the testing using thin ferromagnetic specimen because it has a high conductivity value.

6. Acknowledgement

The authors would like express gratitude to Ministry of Science and Technology and Malaysian Nuclear Agency for the research grant (03-03-01-SF0220) and facilities provided.

References

[1] He Y, Luo F, Pan M, Hu X, Gao J and Liu B 2010 Defect classification based on rectangular pulsed eddy current sensor in different directions *Sensors and Actuators A: Physical.* **157**(2010):26 https://doi.org/10.1016/j.sna.2009.11.012

[2] Tian G Y and Sophian, A 2005 Defect classification using a new feature for pulsed eddy current sensors *NDT & E International.* **38** (2005):77-82 https://doi.org/10.1016/j.ndteint.2004.06.001

[3] Sophian A, Tian, G Y, David T and Rudlin, J 2002 A feature extraction technique based on principal component analysis for pulsed eddy current NDT *NDT & E International.* **36**(2003):37-41 https://doi.org/10.1016/S0963-8695(02)00069-5

[4] Ocal E, M Oral, L E, Erdis E and Vural G 2007 Industry financial ratios—application of factor analysis in Turkish construction industry *Building and Environment.* **42**(2007):385-392 https://doi.org/10.1016/j.buildenv.2005.07.023

[5] Liu C W, Lin K H and Kuo Y M 2003 Application of factor analysis in the assessment of groundwater quality in a black foot disease area in Taiwan *Science of the Total Environment.* **313**(2003) pp 77-89 https://doi.org/10.1016/S0048-9697(02)00683-6

[6] De A, Bandyopadhyay G and Chakraborty B N 2011 Application of the Factor Analysis on the Financial Ratios and Validation of the Results by the Cluster Analysis: *An Empirical Study on the Indian Cement Industry.* **2**(2011) pp 13-31 https://doi.org/10.1063/1.3516362

[7] Cunningham W A, Nezlek J B and Banaji M R 2004 Implicit and explicit ethnocentrism: Revisiting the ideologies of prejudice *Personality and Social Psychology Bulletin.* **30**(2004) pp 1332–1346 https://doi.org/10.1177/0146167204264654

[8] Ramani G and Siegler RS 2008 Promoting broad and stable improvements in low-income children's numerical knowledge through playing number board games. *Child Development.* **79** pp 375–394 https://doi.org/10.1111/j.1467-8624.2007.01131.x

[9] Hair, Jr J F, Black, WC, Babin B J, Anderson R E, Tatham R L 2006 Multivariate data analysis *Pearson-Prentice Hall* (Upper Saddle River, NJ) 6th Ed

[10] Hatcher L 1994 A step-by-step approach to using the SAS(R) system for factor analysis and structural equation modeling (Cary, NC/ SAS Institute)

[11] Kaiser H F 1974 An index of factorial simplicity *Psychometrika* **39**(1974) pp 31–36

[12] Hutcheson G and Sofroniou N 1999 The multivariate social scientist: Introductory Statistics Using Generalized Linear Models *Sage Publication* (Thousand Oaks, CA) https://doi.org/10.4135/9780857028075

IOP Conf. Series: Materials Science and Engineering **554** (2019) 012002 doi:10.1088/1757-899X/554/1/012002

A Guideline of Ultrasonic Inspection on Butt Welded Plates

Ibrahim Burhan[1], Gopinathan Mutaiyah[2], Dina Izzati Hashim[3], Tamil Moli Loganathan[3], Mohamed Thariq Hameed Sultan[4]

[1]Department of Aircraft Maintenance, Politeknik Banting Selangor, Persiaran Ilmu, 42700 Banting, Malaysia
[2]Department of Mechanical Engineering, Universiti Tenaga Nasional, Jalan IkramUniten, 43000 Kajang, Selangor
[3]Department of Mechanical Engineering, Politeknik Banting Selangor, Persiaran Ilmu, 42700 Banting, Malaysia
[4]Department of Aerospace Engineering, Universiti Putra Malaysia, 43400 Serdang, Malaysia

Corresponding author, email: tamilmoli@yahoo.com

Abstract. This paper explained the procedure to carry out the ultrasonic inspection for single and double-V butt welded plates in detail by including the calculation of skip distance, stand-off, interpretation from plotting system, and the preparation of full inspection report as well. In addition, few inspections were compared on butt welded plate single and double-V to determine which is the most suitable, accurate and precise reading obtained. The comparisons were made between scheme answer from Sirim inspector, X-ray film, data that obtained from ultrasonic inspection and calculation by mathematical formulas and using plotting system. Mathematical formulas and plotting system was utilized to obtained the distance of stand-off (S_{off}) and skip distance, (S_{max}) whereas the plotting system were applied to locate the defects based on their S max and S off, and also to determine the depth, d and radius, r. The defects of single-V and double-V butt welding inspected by the ultrasonic technique were then compared with those obtained from the radiographic film while a good agreement was observed. However, the ultrasonic inspection proved to be much more sensitive than the radiographic inspection.

1. Introduction

The ultrasonic inspection method for non-destructive testing is principally based on beams of mechanical waves of short wavelength and high frequency transmitted from a probe and detected by the same or other probes. The attached oscilloscope display with a time base shows the time it takes for an ultrasonic pulse to travel to a reflector (a flaw, the back surface or other free surface) in terms of distance travelled across the oscilloscope screen. The height of the reflected pulse is related to the flaw size as seen from the transmitter probe. The relationship of flaw size, distance and reflectivity are complex, and a considerable skill is required to interpret the data from the display screen [1]. A typical pulse-echo UT inspection system consists of several functional units, such as the pulser/receiver, transducer, and a display device. A pulser/receiver is an electronic device that can produce high voltage electrical pulses. Driven by the pulser, the transducer generates high frequency ultrasonic energy. The sound energy is introduced and propagates through the materials in the form of waves. When there is a discontinuity (such as a crack) in the wave path, part of the energy will be reflected back from the flaw surface. The reflected wave signal is transformed into an electrical signal by the transducer and is displayed on a screen. Knowing the velocity of the waves, travel time can be directly related to the distance that the signal travelled. From the signal, information about the reflector location, size, orientation and other features can be gained [2].

Ultrasonic flaw detection is generally based on a comparative technique. Using appropriate reference standards along with knowledge of sound wave propagation and generally accepted test procedures, a competent operator is able to identify specific echo patterns corresponding to the echo response from good parts and from representative flaws. The echo pattern from a test piece may then be compared to the patterns from these calibration standards to determine its condition [3]. The common defects or flaws found in a welded plate or pipe are lack of fusion (LOF), slag, porosity, lack of penetration (LOP), and cracks. An echo of over penetration easily can be interpreted as a defect, however, only realistic defects or flaws are considered as defects. Therefore, a good understanding of the nature of welding defects is most important before carrying out the inspection. Sometimes, it is difficult to distinguish between side wall crack and slag inclusions. Basically, defects can be categorised as three types as crack like defects (centre crack, side wall crack and LOF), volumetric defect such as porosity and slag and defect in the bottom of weld (LOP and root crack)[4].

2. Experimental Procedure

For ultrasonic inspection, an ultrasonic machine manufactured by Sonatest, with designation transducer of SLC 4-20 (TO 30633), SMA 4-60 (S 2074) and SMA 4-45 (S 2065) was utilized for the inspection. This section explains the inspection of the welded plate using normal probe, angle probe with half and full skip distance, calculation, plotting system and compared with X-ray film.

2.1. Normal Probe

In this inspection method on a parent metal, a normal probe with frequency of 4 MHz was based on the thickness as shown Table 1 below. The parent metal of single-V and double-V welding was inspected for presence of any delamination as illustrated in Figure 1. Delamination may occurs during the manufacturing process due to layer de-bonding or stripping between layers of hot-mix asphalt. This can cause distresses such as longitudinal cracking in the wheel path and tearing in the surface [5].

Table 1. Parent metal thickness, probe and the frequency

Thickness of parent metal (mm)	Type of probe	Frequency (MHz)
Up to 28	Twin crystal	2 to 5
Over 28 to 40	Normal probe	2 to 5

The velocity of the longitudinal waves of steel was fixed to be 5930 mm/s. The test range was calibrated with a test ranges as given in Table 2 or in accordance to the procedure for the plate inspection using V1 Block. The selected test sensitivity for straight beam probe was primary reference echo level (PRE) was set for first back wall echo to 80% full screen height (FSH) from the sound part of the welded plate. The scanning sensitivity and evaluation sensitivity is illustrated in equation (1) and (2). The scanning sensitivity was set to be 23.5 dB for both single-V and double-V welded plate by recording the threshold as 80% of first back wall echo and the range was 100 mm. A masking tape was located along the welding region by covering the welding cap. Hence, any indications of defect can be marked on the masking tape as shown in Figure 1.

The evaluation sensitivity is the PRE value by reducing the gain from scanning sensitivity to PRE value if there is any suspected indication presence during the scanning.

$$\text{Scanning sensitivity} = \text{PRE} + 6\ \text{dB} \tag{1}$$

Evaluation sensitivity = PRE (2)

Table 2. Parent metal thickness and the test range

Parent Metal thickness T (mm)	Test Range (mm)
Up to 22	50
Over 22	100

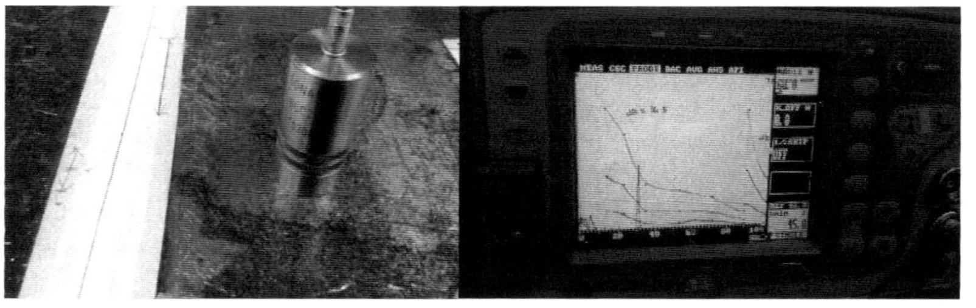

Figure 1. Scanning of parent metal with normal probe

2.2. Angle Probe

The velocity of the shear waves of steel was set to be 3230 mm/s. The angle of beam was selected according to the parent metal thickness as shown in Table 3. IIW or known as V2 block was used for the sweep the range of calibration in addition to ASME basic calibration block was used for the PRE setting. The thickness of the basic calibration block was selected according to the thickness of plate as shown in Table 4. Additionally, the sweep range was calibrated to a test range as indicated in Table 5.

Table 3. Thickness of parent metal, suitable angle and frequency

Parent metal thickness T (mm)	Angle (°)	Frequency (MHz)
Up to 15	60 and 70	2 to 5
Over 15 to 30	45,60 and 70	2 to 5
Over 30 to 40	45 and 70	2 to 5

Table 4. Thickness of parent metal and the block thickness

Parent metal thickness (mm)	Block Thickness
Up to 28	19 mm
Over 28 to 40	38 mm

Table 5. Test range

IOP Conf. Series: Materials Science and Engineering **554** (2019) 012002 doi:10.1088/1757-899X/554/1/012002

Parent Metal Thickness (mm)	Test Range (mm)
Up to 15	100
Over 15 to 40	100-200

Before scanning on welded plate, the Distance Amplitude Correction (DAC) curve should be plotted. DAC provides a means of establishing a graphic reference level sensitivity as a function of sweep distance on the A-scan display [6]. The sensitivity of angle was set by referring to ASME basic calibration block for PRE level as illustrated in Table 3. The threshold was recorded to be 20% DAC. Any indication above 20% of DAC will be evaluated as shown in Figure 2. For 60° angle probe, the scanning sensitivity as in equation (3) and evaluation sensitivity as in equation (4) was set to be 50.5 dB and 44.5 dB respectively for single-V as well as double-V welded plate. However the defect indications were double confirmed by using angle probe 45° by setting the scanning sensitivity and evaluation to 47 dB and 33 dB respectively.

Scanning sensitivity = DAC + 6 dB + Transfer loss (3)
Evaluation sensitivity = PRE + Transfer loss (4)

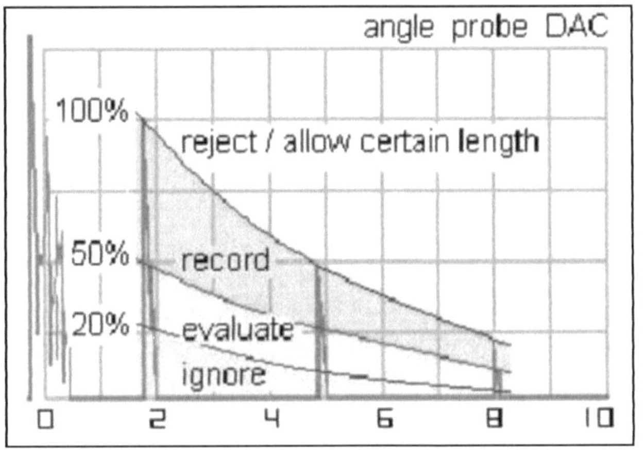

Figure 2. Evaluation and decision making of DAC [7]

2.2.1. Half Skip Distance.
A welded plate as labelled with symbol ^ as shown in Figure 3 indicate where the inspection should begin. The left side is assumed as A(-ve) whereas the right side assumed as B(+ve) region. Half skip distance measured based on equation (5) for critical root scanning by selecting suitable probe according to the thickness of the single-V welded plate. This critical root scanning can be done as lateral scanning along the calculated half skip distance path by using magnetic tape as shown in Figure 4. Nevertheless, for double-V welded plate ¼ skip distance as illustrated in equation (6) was examining as critical root scanning. The couplant a mixture of machine oil and grease was utilized as intermediate liquid between the probe and welded plate.

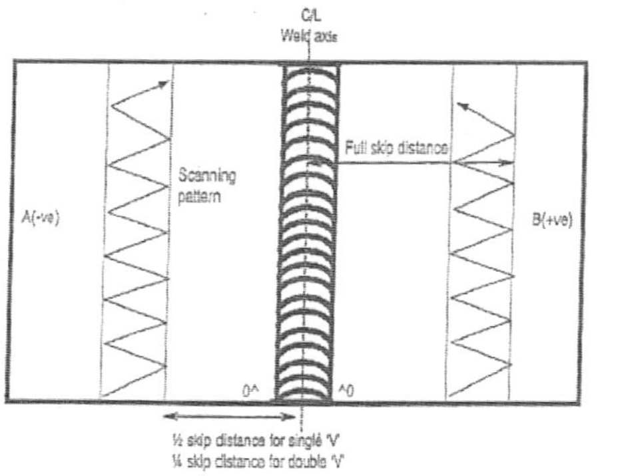

Figure 3. Examination Butt Weld with angle beams.

Figure 4. Lateral scanning for critical root scanning.

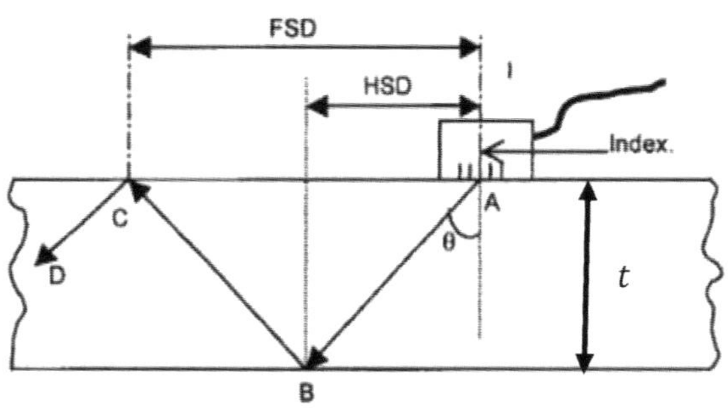

Figure 5. Skip Distance [8]

$$\text{Half skip distance, HSD} = t \times \tan\emptyset \tag{5}$$

$$\text{Quater skip distance, } \tfrac{1}{4}\text{ SD} = \frac{t}{2} \times \tan\emptyset \tag{6}$$

$$\text{Full skip distance, FSD} = 2t \times \tan\emptyset \tag{7}$$

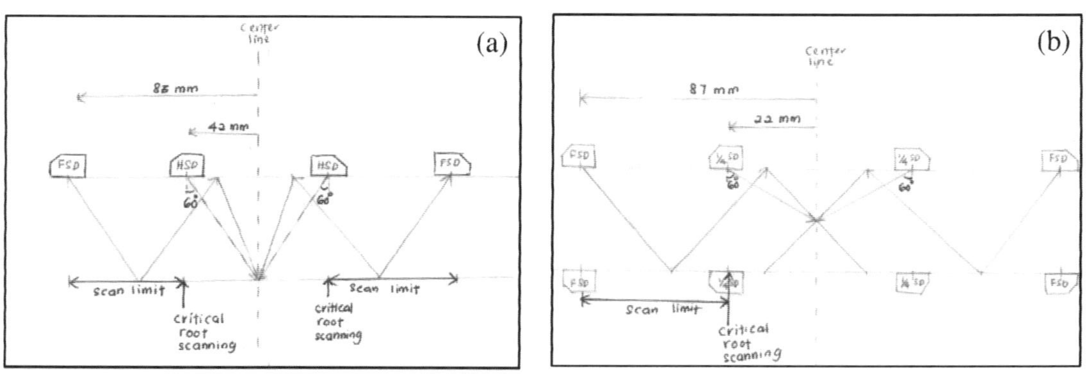

Figure 6. Schemetic of skip distance of probe 60° movement for (a) Single-V and (b) Double-V

2.2.2. Full Skip Distance

Full skip is the distance travelled when the angle beam has travelled down then up as shown in Figure 5. A zig-zag scanning as shown in the Figure 3 is carried between the distances of half or quarter skip distance and full to inspect any defects on the welded region the single-V and double-V welded plates respectively. Inspection on ultrasonic can be on both sides (backside measurement) of double-V welded plate as shown in Figure 6. Table 6 illustrates the type of defects and their description. Table 7 illustrates the scanning pattern for confirming the types of defect that is obtained by observing the amplitude pattern by manipulated with the scanning pattern.

Table 6. Types of defect indication for ultrasonic

Types of defects	Description	HSD FSD	Diagram of echo[9]
Lack of penetration (non-volumetric)	• High amplitude of both side (A and B) of welded plate • Rapidly decreasing in amplitude on rotational scanning	HSD for singleV ¼ SD for double-V	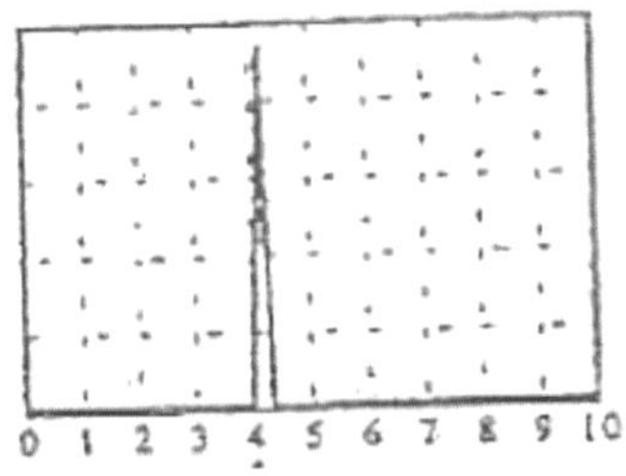
Lack of fusion (non-volumetric)	• High amplitude • Rapidly decreasing in amplitude on the swivel and orbital scanning • Remain the amplitude when lateral	Between HSD and FSD	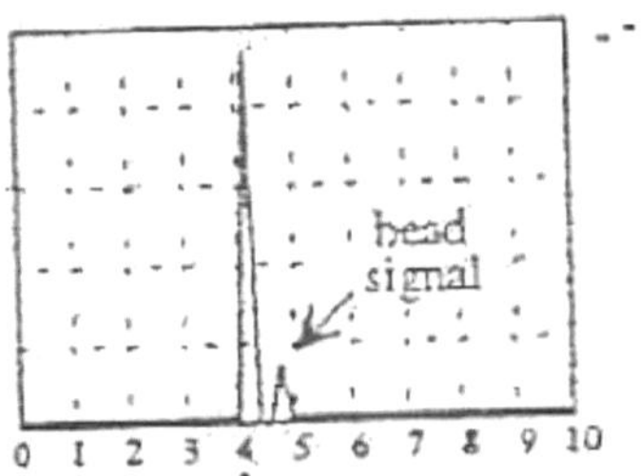
Crack (non-volumetric)	• Crack can be located in toes crack, heat affected zone, center of weld, and at root • High amplitude and multi-faceted reflector	Between HSD and FSD	
	• Signal rise and fall on the swivel and lateral scanning • Decrease in amplitude if orbital scanning		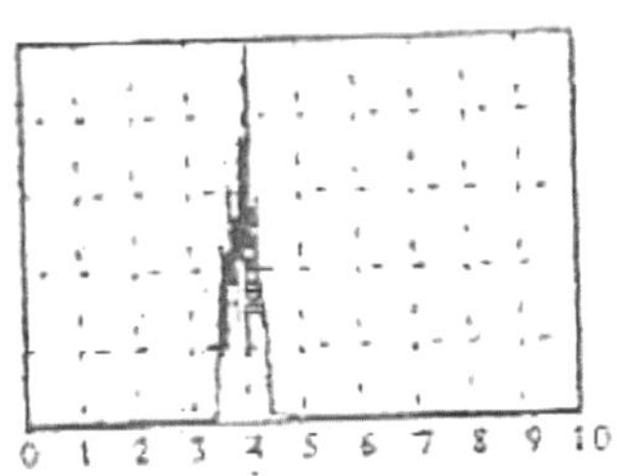

Porosity (volumetric)	• Low amplitude • Giving multiple signals with a wide • Signal maintain on orbital	Between HSD and FSD	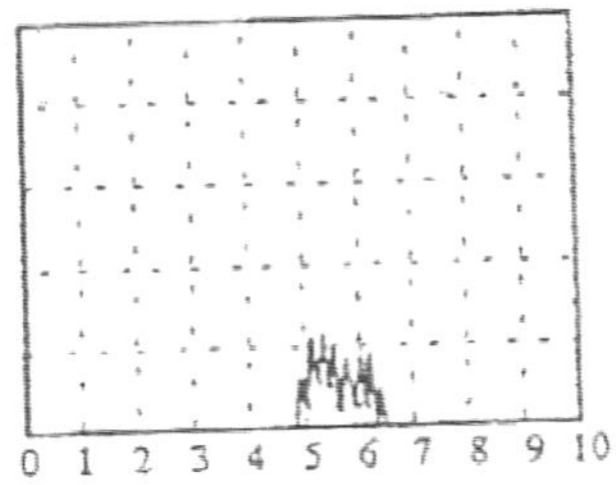
Slag (volumetric)	• Signal contains numerous half cycles and rounded peak • Signal appears to fall as the beck edge rises	Between HSD and FSD	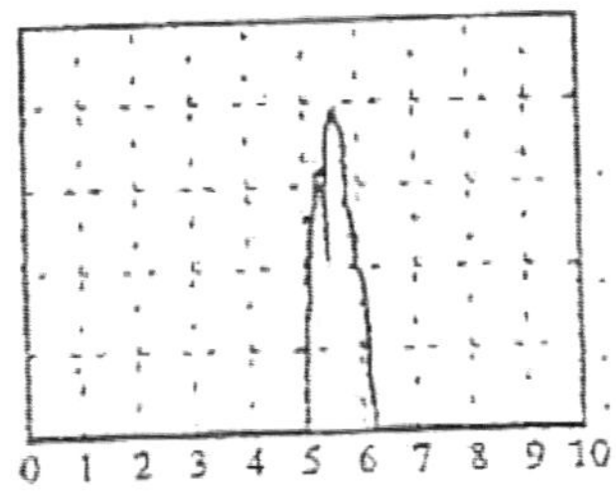

Table 7. Scanning Pattern of Ultrasonic

Scanning patterns	Description	Diagram [9]
Orbital scan	• Probe is manipulated through an arc movement whist maintaining the beam focused on a reflector. • Suitable to identify porosity where the signal is maintained on an orbital scan.	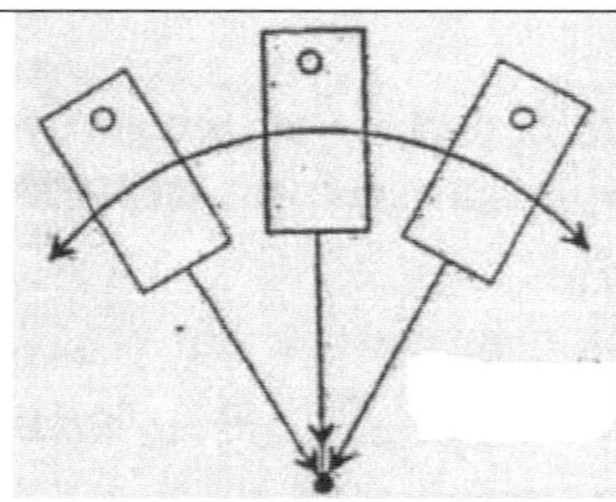
Swivel scan	• Probe is rotated on the spot, effectively scanning the beam around it. • Used to identify multi-faceted, planar or multiple defects	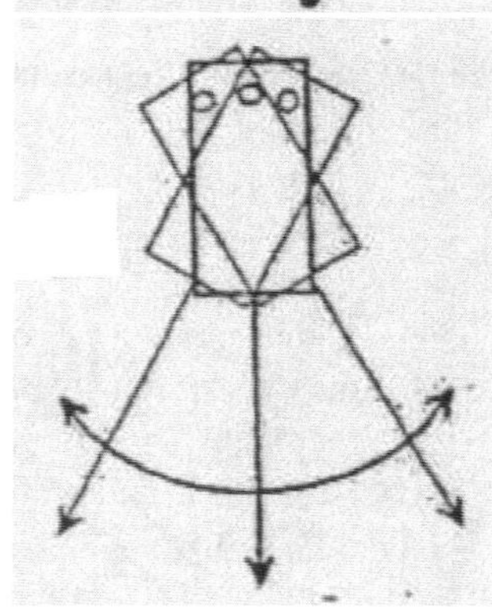

Lateral scan	• Probe is moved sideways along a fixed line. • Used for critical root scanning of single-V weld or for sizing the length of a defect longitudinally.	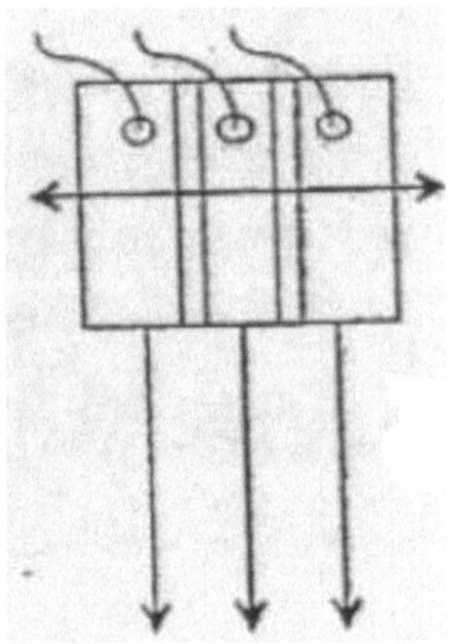
Depth scan	• Probe is moved back and forth in the direction of the beam. • As in locating the position of a defect when maximising the signal off a transverse hole to set sensitivity.	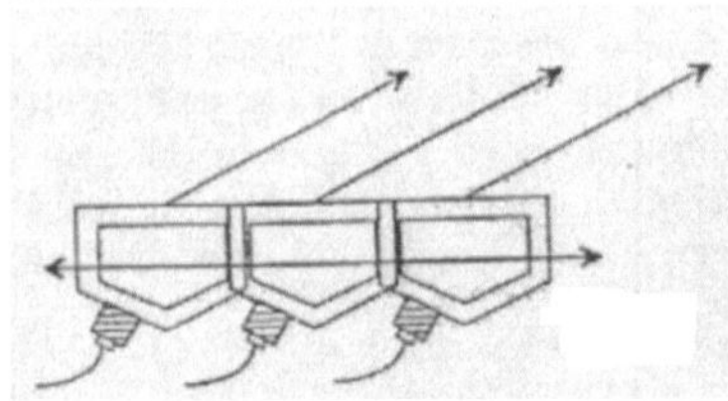

2.2.3. By Calculation

When performing an angle beam inspection, it is important to know where the sound beam is encountering an interface and reflecting. The reflection points are sometimes referred to as nodes. The location of the nodes can be obtained by using the trigonometric functions or by using the trigonometric based formulas as shown in equation (8) and (9). Stand-off is referring the distance of probe index to the center line, however surface distance or known as skip distance is the distance from the probe index where the sound beam returns to the surface.

Stand-off of half skip, $S_{off} = t \times \tan Ø$ (8)

Surface distance, $S_{max} = t / \cos Ø$ (9)

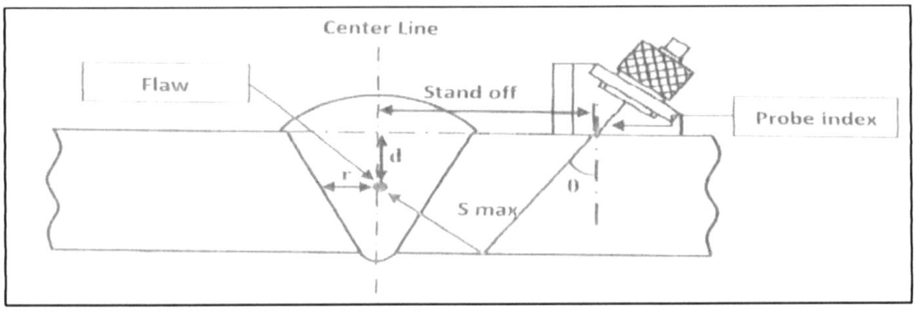

Figure 7. Location of probe

2.2.4. Plotting Card

In angle probe scanning plotting systems are used for projecting defect depths and positions in relation to the probe index by applying the beam path, read from the screen (CRT time base) and the stand-off or surface distance from a reference datum on the welded plate. By drawing probe angle onto a plotting card, while overlapping the reflector welding portion on the plotting card based on standoff and surface

IOP Conf. Series: Materials Science and Engineering **554** (2019) 012002 doi:10.1088/1757-899X/554/1/012002

distance to indicate the position of the defects as shown in Figure 8. The exact locations labelled as 1, 2 and 3 are supposed to reflect on the mirror line as shown in Figure 9 and 10.

Once the locations in the tracing paper as in Figure 9 and 10 are marked, Figure 11 should be filled up. Figure 11 describes the angle probe that are used, types of defects that are encountered, value of dB above the 100% of DAC curve, value of stand-off, and maximum surface distance. The description of sizing of the defects as shown in Figure 7 are as below:

L = length of the defects
d = depth of the defects from upper surface
r= radius of the defect from centre line (-ve for left and +ve for right)
q = length from datum as marked as ^ as shown in Figure 3

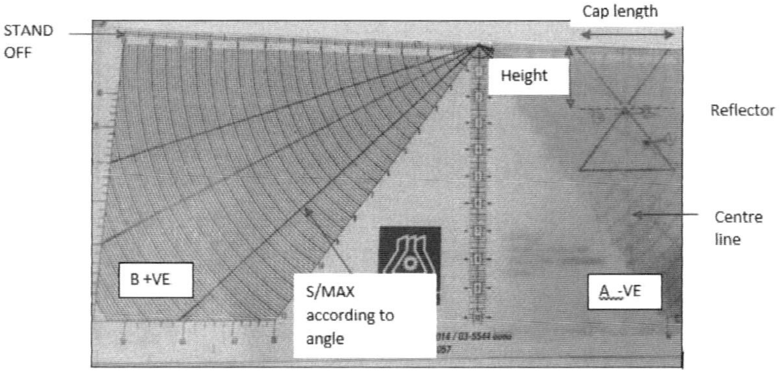

Figure 8. Plotting Card

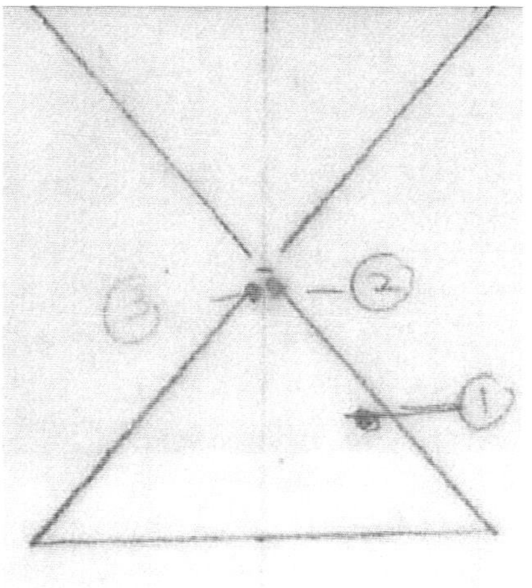

Figure 9. Location of defects in Single V.

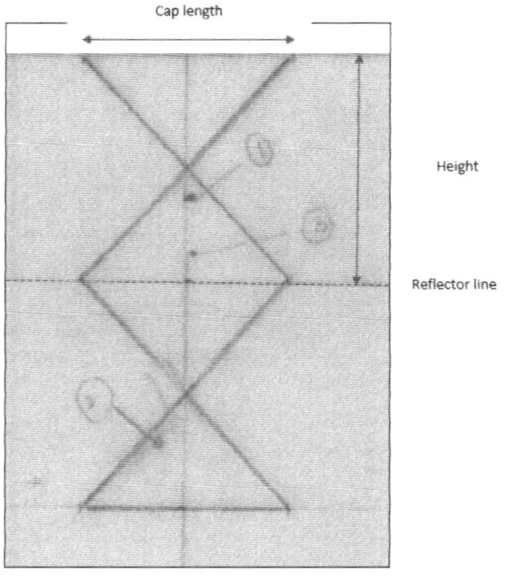

Figure 10. Location of defects in Double V.

IOP Conf. Series: Materials Science and Engineering **554** (2019) 012002 doi:10.1088/1757-899X/554/1/012002

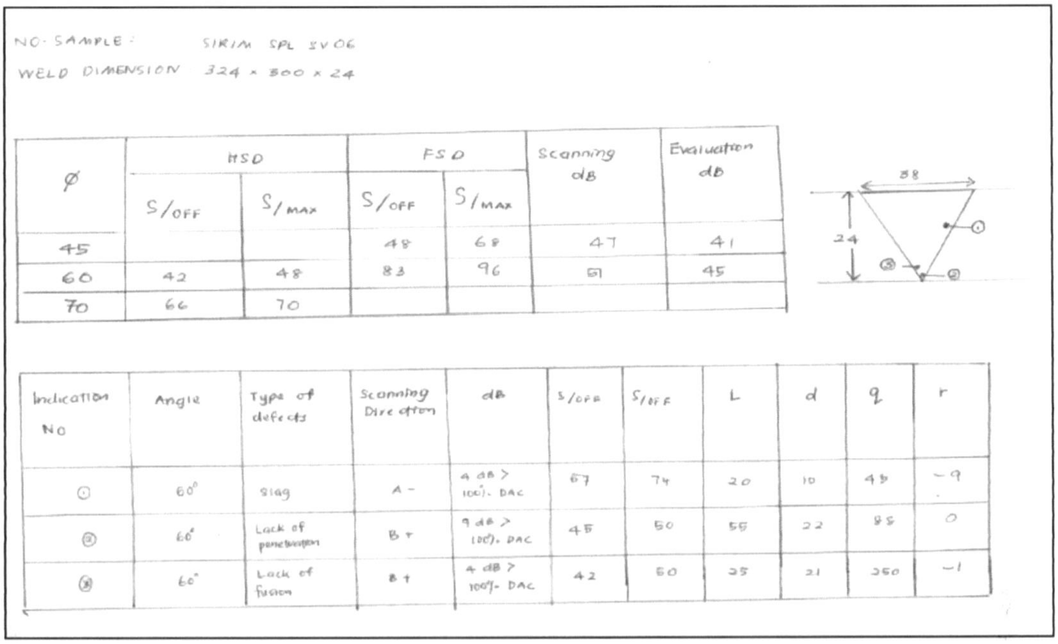
NO. SAMPLE : SIRIM SPL SV 06
WELD DIMENSION 324 x 300 x 24

ø	HSD		FSD		Scanning dB	Evaluation dB
	S/OFF	S/MAX	S/OFF	S/MAX		
45			48	68	47	41
60	42	48	83	96	51	45
70	66	70				

Indication No	Angle	Type of defects	Scanning Direction	dB	S/OFF	S/OFF	L	d	q	r
①	60°	slag	A −	4 dB > 100% DAC	67	74	20	10	43	−9
②	60°	Lack of penetration	B +	9 dB > 100% DAC	45	50	55	22	85	0
③	60°	Lack of fusion	B +	4 dB > 100% DAC	42	50	25	21	250	−1

Figure 11. Recording the data in the table

2.2.5. X-ray

An X-ray film in SPL SV 06 was viewed using film viewer and the defects are clearly seen by switching of the power and sizing could be done using ruler to measure the length of the defect, L and from datum, 0.

Figure 12. Result of X-ray for (a) Single-V and (b) Double-V

IOP Conf. Series: Materials Science and Engineering **554** (2019) 012002 doi:10.1088/1757-899X/554/1/012002

3. Result and Discussion

Table 8 shows the signal of defects for single-V that was obtained from SIRIM scheme by inspectors and compared with signal that was obtained by ultrasonic inspection. Basically signal that obtained by both parties are morally the same. It is also important to note that when using shear waves to know where along the probe the beam enters probe index. The probe index position relative to some datum on the specimen and the exact beam angle allows to calculating the horizontal and vertical distances.

Table 8. Defect Signal of Single V

Location	**1**	**2**	**3**
Type Scheme Sirim	Slag 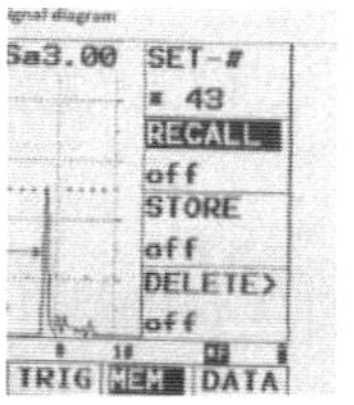	LOP 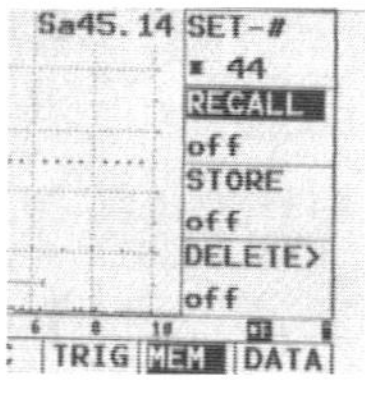	LOF 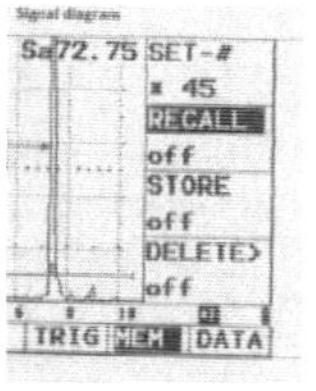
Location	**1**	**2**	**3**
Type Ultrasonic Inspection	Slag 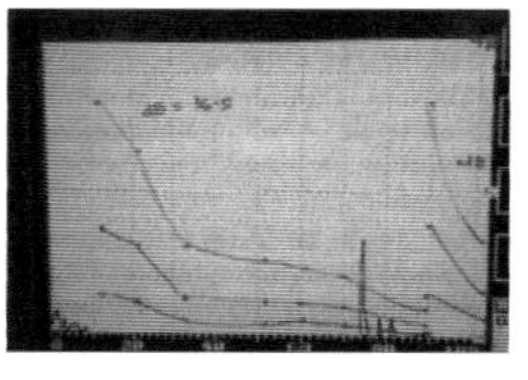	LOF 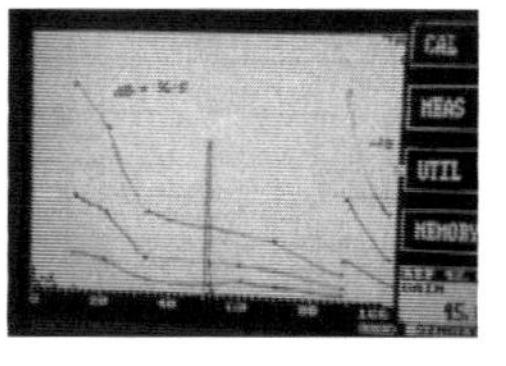	LOP 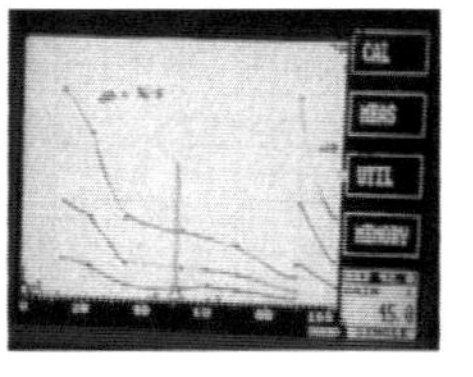

Table 9 shows the defects location and the sizing by comparison which is made from SIRIM scheme with ultrasonic inspection. In this case, both parties are using angle probe, 60°, however for defect no 2 and 3, though we used opposite direction of scanning, it did not influence the accuracy of reading that was obtained. The data of sizing basically same and do not have much significant. The difference in length of scheme was 42% while ultrasonic inspection was 4% if compared with result of X-ray. In general there are three techniques that can be utilized for sizing the defect such as applying maximum amplitude, 6 dB drop or 20 dB drop. For ultrasonic inspection of these welded plate sizing was based on 20 dB drop was used throughout the inspection. However, all these three techniques should give the same reading. However, the hand pressure on the probe might influence the reading as well.

Table 9. Diagram and sizing of defects in Single-V welded plate

	Scheme of SIRIM	**Ultrasonic inspection**
Diagram	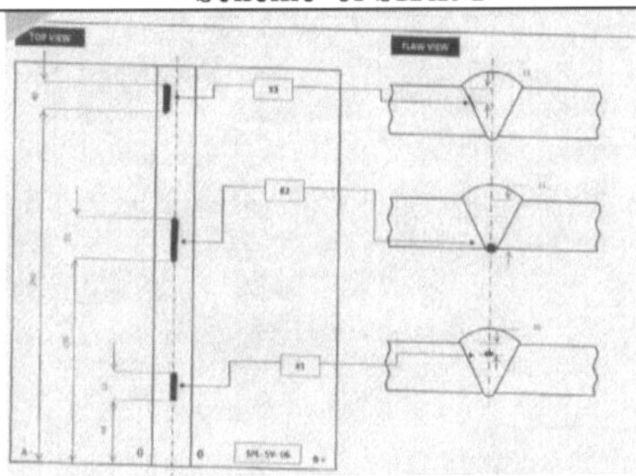	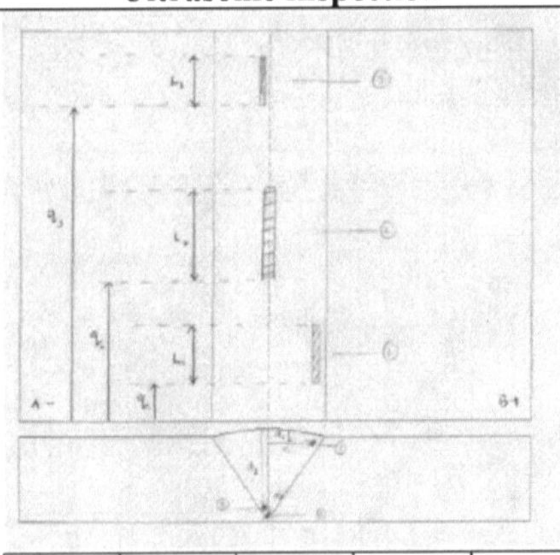

Sizing

Scheme of SIRIM:

No.	L	q	d	r
1	12	46	11	7
2	51	100	22	0
3	40	240	10	0

Ultrasonic inspection:

No.	L	q	d	r
1	20	45	10	9
2	55	88	22	0
3	25	250	21	-1

On the other hand, Table 10 shows the signal of defects for double-V that was obtained from SIRIM scheme by inspectors and compared with signal that was obtained by ultrasonic inspection. Basically signal that obtained by both parties are morally the same.

Table 10. Defect Signal of Double-V

Location	**1**	**2**	**3**
Type Double Plate	Slag 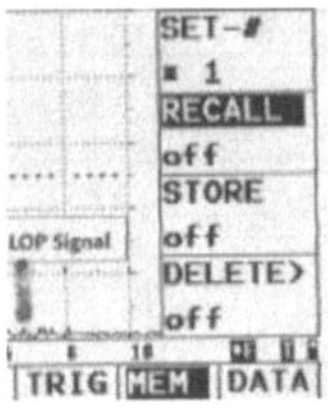	LOF 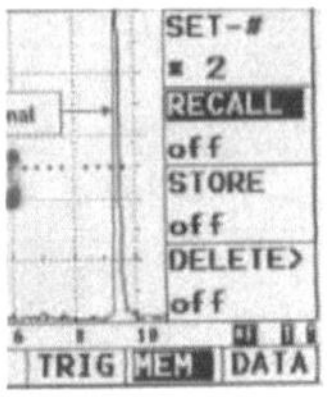	LOP 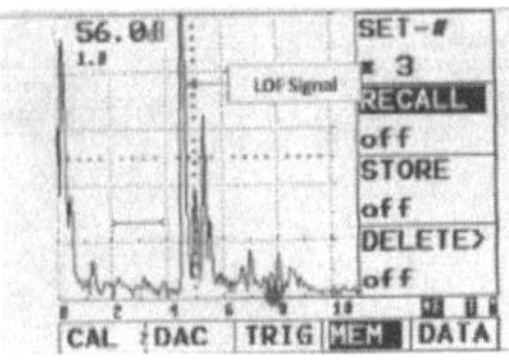
Location	**1**	**2**	**3**
Type Double Plate	Slag 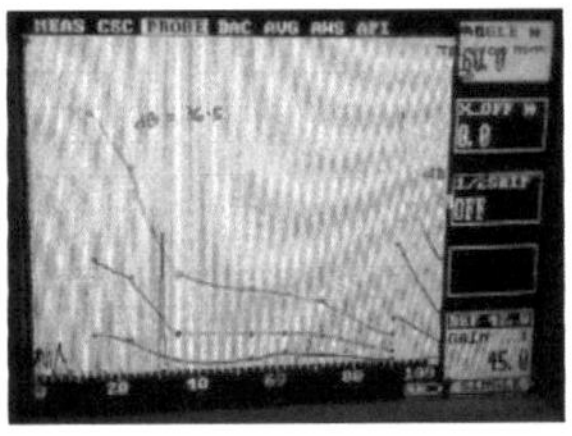	LOF 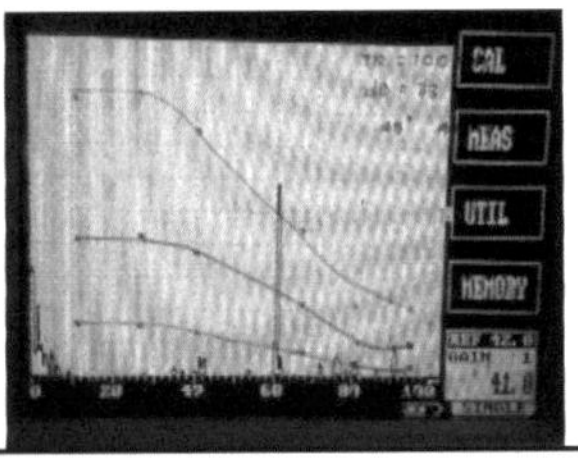	LOP 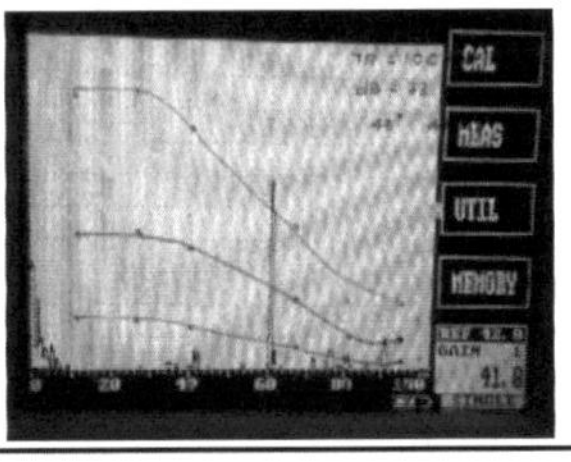

Table 11. Diagram and sizing of defects in Double-V welded plate

	Scheme of Sirim	Ultrasonic inspection
Diagram	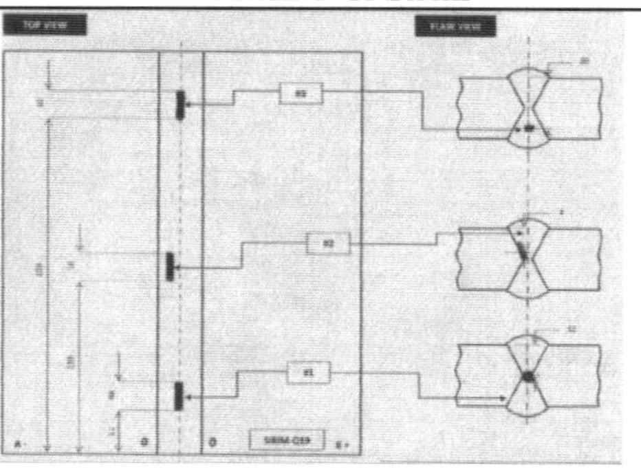	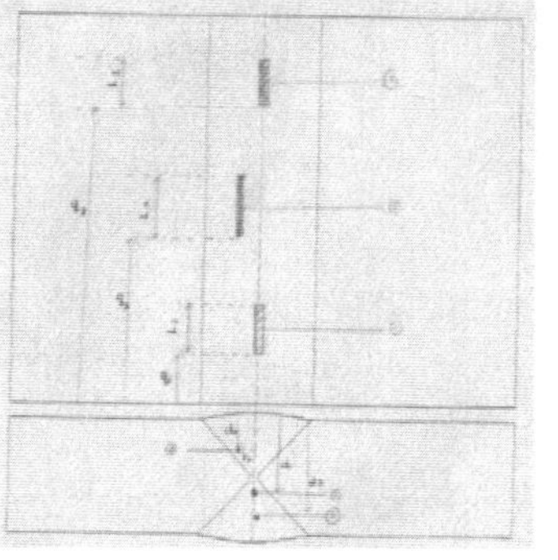

Sizing

Scheme of Sirim:

No.	Flaw type	Angle, Ø	Scanning area
1	Slag	60	A-
2	LOF	60	B+
3	LOP	60	B+

No.	l	q	d	r
1	11	48	12	0
2	16	139	4	6
3	10	259	20	0

Ultrasonic inspection:

No.	Flaw type	Angle, Ø	Scanning area
1	Slag	60	A-
2	LOF	45	A-
3	LOP	60	B+

No.	l	q	d	r
1	20	65	15	0
2	20	143	8	-3
3	17	250	22	1

Similarly, defect length is usually a significant criteria for accept/reject of a weld. Since most codes acceptance criteria for ultrasonic inspection are conversions of radiographic acceptance criteria, the defect lengths are based on workmanship, therefore they are conservative. Typically 10-12mm would be the maximum allowable length for some flaw types. However, for short flaws, the normal dB drop method measures only beam width not defect length.

Table 12. A comparison between few techniques

	Single-V			Double-V		
	Defect 1	Defect 2	Defect 3	Defect 1	Defect 2	Defect 3
	Slag	LOP	LOF	Slag	LOF	LOP
	Length (mm)			Length (mm)		
Data ultrasonic inspection	20	55	25	20	20	17
Scheme of SIRIM	12	57	40	11	16	10
X-ray	21	54	Not detectable	25	20	10

Scheme from SIRIM	12	51	40	11	16	10

A comparison of data that was obtained from X-ray, ultrasonic inspection, calculation by mathematical equation, and plotting system was carried out. Based on the Table 8 and Table 9 for LOP and LOF giving maximum amplitude due to the non-volumetric region. Based on X-Ray result on butt welded single-V film, the lack of fusion was not clearly visible. It may be due to the angle or direction of the X-ray source. X-ray is supposed to be carried at ± 30° to detect side wall fusion which is a volumetric defect. Therefor X-ray should be taken at few angles to obtain an accurate reading. Indeed ultrasonics are very sensitive to these types of defects. This shows that an ultrasonic test is most suitable inspection to detect the defect by covering all the region of welding by angle beam, even though it's consuming time compared to X-ray. Conversely, in X-ray examination there is potential to miss defects.

4. Conclusion

Defect sizing is very essential for performing structural integrity of a material. The defects of single-V and double-V butt welding inspected by the ultrasonic technique were then compared with those obtained from the radiographic film. A good agreement was observed. It exists in different ultrasonic sizing techniques based on signal amplitude but the data cannot be stored. All the indication will appear on the display so that instant interpretation have to be done. Ultrasonic is much more sensitive but it requires a high degree of skill in interpreting pulse-echo patterns. However, a permanent record is not readily obtained when compared with radiography. On the other hand, radiography defect indication is recorded on film hence giving a permanent record. However, skill is needed for selecting the best angles of exposure and interpreting indications. At the same time, strict safety precautions must be observed. It's not generally suitable for fillet welding inspection.

Acknowledgment

The authors would like to express their gratitude to the Management of Politeknik Banting Selangor (PBS), Centre of Technology (NDT), PBS and SIRIM Berhad officer especially from NDT Department, who provided insight and expertise that greatly assisted in this research by providing the facilities and equipment.

Reference

[1] *Ultrasonic flaw detection*. Available: http://www.bindt.org/What-is-NDT/Ultrasonic-flawdetection/

[2] Introduction to Non-Destructive Testing Techniques.

[3] T. Nelligan. *Ultrasonic Flaw Detection* Available: http://www.olympusims.com/en/applications-and-solutions/introductory-ultrasonics/introduction-flaw-detection/ [4]

[4] F. Lingvall and T. Stepinski, "Ultrasonic Characterization of Defects," 1999.

[5] K. Hoegh, L. Khazanovich, K. Maser, and N. Tran, "Evaluation of Ultrasonic Technique for Detecting Delamination in Asphalt Pavements," *Transportation Research Record: Journal of the Transportation Research Board,* vol. 2306, pp. 105-110, 2012.

[6] *NDT Resource Center* Available: https://www.ndeed.org/EducationResources/CommunityCollege/Ultrasonics/CalibrationMeth/DAC_Curve.ht m

[7] Jayesh. (2011). *Ultrasonic*. Available: https://www.scribd.com/doc/66853983/Ultrasonic

IOP Conf. Series: Materials Science and Engineering **554** (2019) 012002 doi:10.1088/1757-899X/554/1/012002

[8] *Ultrasonic Examination Part 1*. Available: http://www.twi-global.com/technicalknowledge/job-knowledge/ultrasonic-examination-part-1-127/
[9] *Ultrasonic Inspection* Issue 10 24/05/11). Ruane TATI Sdn Bhd.

IOP Conf. Series: Materials Science and Engineering **554** (2019) 012003 doi:10.1088/1757-899X/554/1/012003

A Millimetre-wave Probe for Fast Screening and Evaluation of Corrosion in Planar Conductors: Numerical Simulations

Wenjia Li[1], Yong Li[1,*], Jianguo Tan[1], Ilham Mukriz Zainal Abidin[2], Jinhua Hu[1], Zhenmao Chen[1], Yongzhuo Wu[3]

[1] State Key Laboratory for Strength and Vibration of Mechanical Structures, Shaanxi Engineering Research Centre of NDT and Structural Integrity Evaluation, School of Aerospace Engineering, Xi'an Jiaotong University, Xi'an, Shaanxi, China
[2] Leading Edge NDT Technology (LENDT) Group, Malaysian Nuclear Agency, 43000 Bangi, Kajang, Selangor, Malaysia
[3] Jiangsu Eqotek Testing Services Co. Ltd, Kunshan, Jiangsu, China

Corresponding author, email: yong.li@mail.xjtu.edu.cn

Abstract. Planar conductors are commonly used in critical mechanical structures. Due to high-temperature and high-pressure environment, corrosion may occur in in-service planar conductors, and seriously threatens structural integrity and safety. Hence, it is indispensable to detect and evaluate corrosion using Non-destructive Testing (NDT) techniques before structural failure. As a technique complementary to other Electromagnetic NDT methods, Microwave NDT is found to be advantageous in terms of high efficiency and accuracy in inspection of conductors subject to surface flaws. In this paper, a millimetre-wave probe for efficient detection and evaluation of corrosion in planar conductors is proposed and intensively investigated via simulation. A 3D Finite Element (FE) model of the proposed probe is established based on electromagnetic wave theory. The characteristics of the microwave propagation along with the mechanism regarding corrosion detection are intensively analysed. The probe structure and parameters are subsequently optimised in an effort to enhance the detectability, sensitivity and accuracy in corrosion evaluation. Besides, signal processing techniques involving extraction of features from the S-parameter in the function of frequency are investigated.

1. Introduction

Planar conductors are commonly used in critical mechanical structures. Due to high-temperature and high-pressure environment, corrosion may occur in in-service planar conductors, and seriously threatens structural integrity and safety. Hence, periodic non-destructive testing for planar conductors in large complex structures plays a key role to guarantee the safety and reliability of the structures. Conventional non-destructive testing methods mainly include Radiographic Testing (RT) [1], Eddy Current Testing (ECT) [1] and Ultrasonic Testing (UT) [2]. RT may be pernicious to inspectors, and thus protective measure must be taken. ECT is scarcely efficient and susceptible to interference from the material itself and other factors. UT requires coupling agent deployed over the surface of the specimen, and is merely applicable for the detection of subsurface defects. As a complementary technique to other Electromagnetic Non-destructive Testing (NDT) methods, Microwave NDT has the advantage over conventional methods in efficiency and sensitivity [3].

IOP Conf. Series: Materials Science and Engineering **554** (2019) 012003 doi:10.1088/1757-899X/554/1/012003

A typical Microwave NDT system is normally comprised of a microwave generator, coaxial cables and probes. The generator is used for generating the microwave, the coaxial cables are the connector between the generator and the probes, and the probes are the key components for transmitting and receiving the microwave [4]. The probes used for microwave NDT generally include horn antennas [5], waveguides [6] and coaxial line sensors [7,8,9]. Since the propagation direction of the microwave transmitted by the probe is perpendicular to the specimen surface, the probe should be scanned over the specimen surface with small steps to achieve high-spatial-resolution inspection [10]. This makes it inefficient to use these probes for Microwave NDT. Therefore, it is highly desirable to develop a new type of Microwave NDT probe with high detection efficiency and sensitivity.

This paper proposed a new probe working in K-band (i.e. a millimetre-wave probe) for detection and evaluation of corrosion in planar conductors. The probe consists of a rectangular metal box in which part of the bottom surface is in absence, and an excitation port (Coaxial Port #1) and a detection port (Coaxial Port #2) both embedded on the upper surface of the metal box. The probe ports are connected with the vector network analyser for microwave excitation and acquisition. Figure 1 illustrates the schematic diagram of the Microwave NDT system together with the cross-sectional view (along *y*-axis) of the proposed probe. The detection principle underlying the proposed probe is that the excited microwave horizontally propagates in the waveguide structure which is composed of the probe metal box and the upper surface of the planar conductor under inspection. The corrosion occurring in the conductor surface perturbs the transmission of the microwave from the excitation port to detection port. This results in the variations in: (1) the transmitted microwave received by the detection port; and (2) the reflected wave captured by the excitation port. Compared with the scenario for flawless conductors, the resulting signals, which are the S-parameters (i.e. S_{11} and S_{21} [11]) in the function of excitation frequency, consequently vary with the presence of corrosion, giving implication regarding corrosion parameters.

In the rest of this paper, the intrinsic mechanism regarding the detection and evaluation of corrosion by using the proposed probe is analysed via Finite Element (FE) simulations. Based on this, the probe is further optimised particularly regarding the matching port, metal ring of the coaxial port and detection mode. The correlations of signals of S_{11} and S_{21} and their features with the corrosion parameters are subsequently investigated.

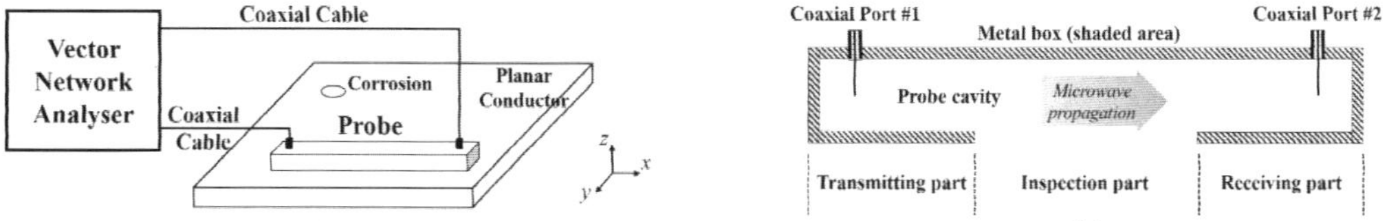

Figure 1. (a) Schematic diagram of the system along with the proposed millimetre-wave probe. (b) Cross-sectional illustration of the proposed probe.

2. Numerical simulations

In a bid to investigate the characteristics of millimetre-wave propagating in the probe and optimisation of probe parameters, a 3D Finite Element (FE) model has been built up based on finite element modelling. A series of FE simulations are intentionally conducted subsequently. It should be noted that the model is solved with approximately 835,164 tetrahedral elements.

2.1. Model setup.

2.1.1. The wave equation and material settings.

IOP Conf. Series: Materials Science and Engineering **554** (2019) 012003 doi:10.1088/1757-899X/554/1/012003

The model with the proposed probe deployed over a planar conductor is illustrated in Figure 2(a). In the simulation, the field quantity i.e. electric field intensity **E** is obtained by numerically solving the wave equation of microwave propagation in the air, dielectric material and conductor. The wave equation is written as [12]:

$$\nabla \times \mu_r^{-1}(\nabla \times \mathbf{E}) - k_0^2\left(\varepsilon_r - \frac{j\sigma}{\omega\varepsilon_0}\right)\mathbf{E} = 0 \tag{1}$$

where, μ_0 and ε_0 represent the permeability and dielectric constant of vacuum, respectively. μ_r, ε_r and σ indicate the relative permeability, relative dielectric constant and electrical conductivity of a material in the model. ω denotes the angular frequency. k_0 is the wave number.

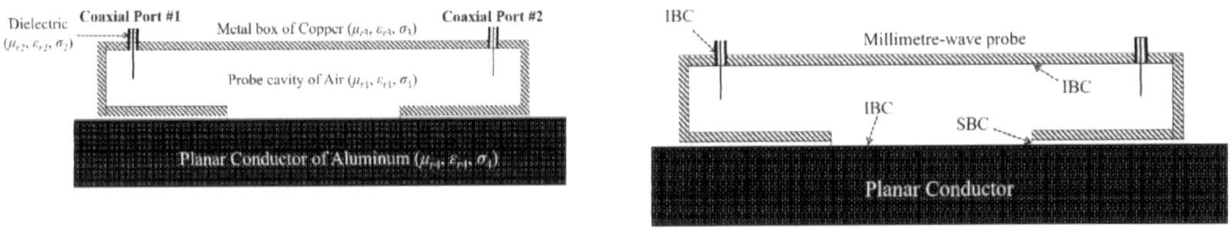

Figure 2. (a) The material for each region by cross-sectional illustration. (b) Setup of the boundary conditions.

In Figure 2(a), the material in the probe cavity is air whilst the material of the coaxial port region is dielectric. The material of the metal box of the probe is copper whilst the planar conductor under inspection is an aluminium slab. The material properties of the air, dielectric material, copper and aluminium involved in simulations are listed in Table 1.

Table 1. Properties of the materials involved in simulations.

Material	Property	Name	Value
Air	Relative permeability	μ_{r1}	1
	Dielectric constant	ε_{r1}	1
	Electrical conductivity	σ_1	0 S/m
Dielectric Material	Relative permeability	μ_{r2}	1
	Dielectric constant	ε_{r2}	2
	Electrical conductivity	σ_2	0 S/m
Copper	Relative permeability	μ_{r3}	1
	Dielectric constant	ε_{r3}	1
	Electrical conductivity	σ_3	59.98 MS/m
Aluminium 6063-T83	Relative permeability	μ_{r4}	1
	Dielectric constant	ε_{r4}	1
	Electrical conductivity	σ_4	30.30 MS/m

2.1.2. Boundary conditions.

The behaviour of the microwave at the interface between 2 different materials is important for analysis of electromagnetic phenomena, and constituted by boundary conditions. The setup of the boundary conditions regarding the FE model is illustrated in Figure 2(b). In the model, the Impedance Boundary Condition (IBC) is firstly applied to the interfaces between the metal and dielectric material in light of the fact that there is impedance loss at the interface.

It is noted that the use of IBC is based on the assumption that the skin depth of the electromagnetic field in the conductor under microwave illumination is much less than the conductor thickness. Therefore, IBC can only be applied to the surface of the conductor with the thickness considerably larger than the skin depth. The IBC is written as [12]:

IOP Conf. Series: Materials Science and Engineering **554** (2019) 012003 doi:10.1088/1757-899X/554/1/012003

$$\sqrt{\frac{\mu_0\mu_r}{\varepsilon_c}}\mathbf{n}\times\mathbf{H}+\mathbf{E}-(\mathbf{n}\cdot\mathbf{E})\mathbf{n}=(\mathbf{n}\cdot\mathbf{E}_S)\mathbf{n}-\mathbf{E}_S \quad (2)$$

where, the source electric field $\mathbf{E}_S$ can be used to specify a source surface current on the boundary.
In addition to the boundary conditions between the metal and the media, there is a certain gap between the probe and the planar conductor (namely probe lift-off), which causes the leakage of microwave. Since only the distribution of electromagnetic wave in the waveguide structure composed of the probe metal box and planar conductor is concerned, the Scattering Boundary Condition (SBC) directly absorbs the microwaves leaking from the outward radiation is applied, which is expressed as [12]:

$$\mathbf{E}=\mathbf{E}_{sc}e^{-ik(\mathbf{n}\cdot\mathbf{r})}+\mathbf{E}_0e^{-ik(\mathbf{k}\cdot\mathbf{r})} \quad (3)$$

where, the field $\mathbf{E}_0$ is the incident plane wave that travels in the direction **k**.

With the consideration of the wave equation for microwave propagation and the boundary conditions, the model is thus completely established, and subsequently solved via FEM.

2.2. Probe optimisation.

2.2.1. Optimisation regarding the port matching.

Simulations have been firstly conducted in a bid to optimise the structure of the transmitting/receiving part of the probe for efficient transmission of microwave from the coaxial port to the inspection part. Note that as indicated in Figure 1(b), the transmitting/receiving part of the probe consists of a coaxial port and cuboid similar to waveguide structure. Three types of the structures including the rectangular port, open port and quarter-wavelength waveguide are investigated in simulations, and portrayed in Figure 3. For each structure, S_{11} depicting the energy of the microwave reflected to the excitation port is simulated and analysed. The simulation results are presented in Figure 4.

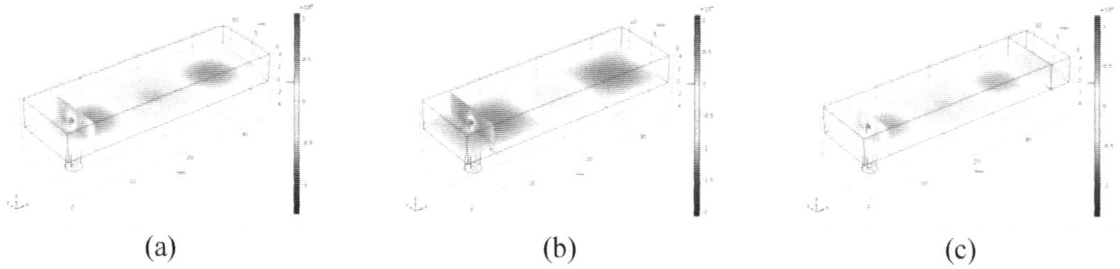

(a) (b) (c)

Figure 3. Structures of the transmitting part of the proposed probe: (a) The rectangular port; (b) Open Circuit port; and (c) Quarter-wavelength waveguide.

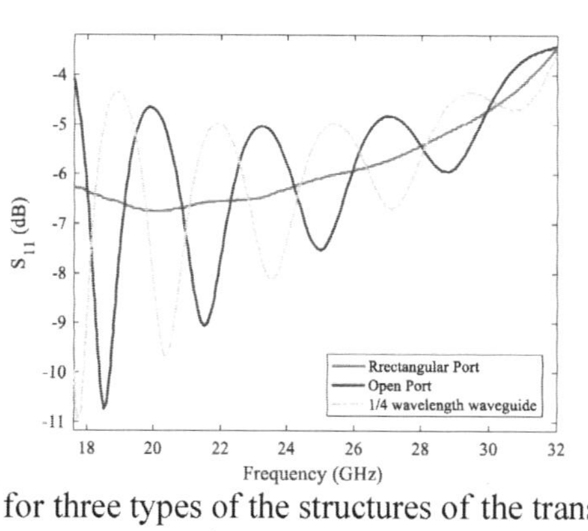

Figure 4. S_{11} *vs.* frequency for three types of the structures of the transmitting part: the rectangular port, open port and quarter-wavelength waveguide

It can be found from Figure 4 that for the entire frequency range, compared with the significant fluctuation in S_{11} for the open-circuit matching structure and the quarter-wavelength waveguide, the value of S_{11} for the rectangular port is relatively stable. This reveals that the rectangular port is stable over the frequency band of the proposed probe, and gives relatively low S_{11} value. This can be taken as the preferred matching option. The simulation result indicting the field distribution along with the energy flow of the microwave propagating through the rectangular-port-based transmitting part of the probe is presented in Figure 5. It can be seen that the electric field intensity distribution in the transmitting part of the probe is uniform, and a large amount of microwave energy in the transmitting part of the probe is transmitted to the detection area. The designed metal ring plays an indispensable role in the efficient excitation and transmission of the microwave.

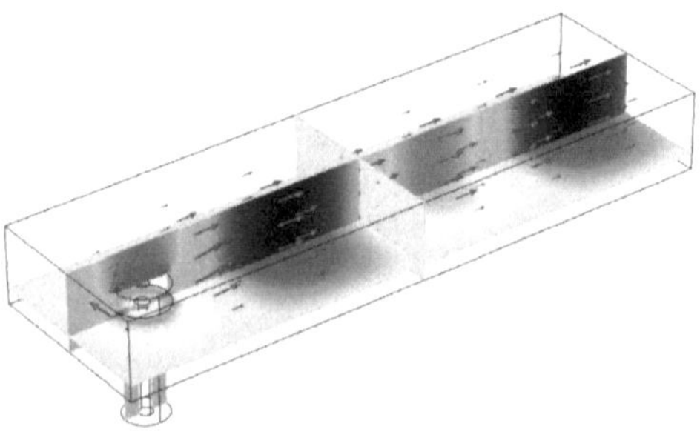

Figure 5. Distributions of the electric field and energy flow of the microwave propagating in the probe cavity.

2.2.2. Optimisation of the metal ring.

In an effort to improve and probe excitation performance and enhance the transmission of microwave, a metal ring is usually introduced to the coaxial port of a microwave NDT probe. As a result, the metal ring added to each coaxial port of the proposed probe is taken into account in the model. The size of the metal ring in terms of l_{mr} for the height of the metal ring and r_{mr} for the ring radius are thus optimised by S_{11} acquired from simulations. The sizes of the metal ring and the corresponding S_{11} against different excitation frequencies are listed in Table 2.

Table 2. The size of metal ring and S_{11} *vs.* excitation frequency

Excitation frequency (GHz)	l_{mr} (mm)	r_{mr} (mm)	S_{11} (dB)
20	1.30	0.95	-45
21	0.95	2.15	-33
22	0.70	1.60	-49.6
23	0.85	1.10	-39
24	0.95	1.55	-40
25	0.60	1.60	-40

From Table 2, it can be seen that S_{11} at the excitation frequency of 22GHz is minimum when the metal ring size is 0.7mm×1.6mm (l_{mr}×r_{mr}). This indicates that such metal ring size corresponding to -49.6dB could be preferable. Whereas, in consideration of the fabrication and centre frequency of the operation band of the probe (approx. 23GHz), the preferred size of the metal ring is 0.85mm×1.10mm.

2.3. Characteristics of the microwave propagation in the probe.

After the optimisation of the critical parts of the probe, the microwave propagation within the probe cavity is simulated. The distribution and electric field and power flow of the microwave are investigated together with the simulated S_{11} and S_{21} which are acquired at Coaxial Port #1 and Coaxial Port #2, respectively. The simulation results are presented in Figure 6.

IOP Conf. Series: Materials Science and Engineering **554** (2019) 012003 doi:10.1088/1757-899X/554/1/012003

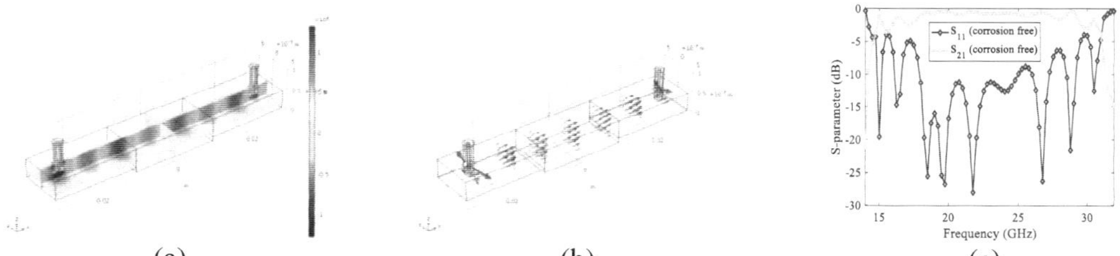

(a) (b) (c)

Figure 6. Simulation results: (a) *z* component of electric field intensity in the probe; (b) power flow inside the probe; and (c) S-parameters for the probe placed on the flawless conductor.

From Figure 6(a)-(b), it can be seen that a large majority of the microwave energy is transmitted from Coaxial Port #1 to Coaxial Port #2. This facilitates the detection and evaluation of corrosion in the planar conductor which is illuminated by incident wave from the inspection part of the proposed probe. It is also noticeable from Figure 6(c) that when the conductor is free of corrosion, the amplitude of S_{11} is considerably higher than that of S_{21}. This implies the high efficiency of the microwave propagating through the probe. Since S_{11} and S_{21} could vary in the presence of corrosion, they are subsequently utilised for corrosion evaluation.

3. Detection and evaluation of corrosion using the proposed probe

After the design and optimisation of the probe, FEM simulations are conducted for investigation of corrosion evaluation based on signals of S_{11} and S_{21}. Corrosion in the cylindrical shape with variable depths and radii are taken into account in simulations.

3.1. Corrosion in different depths

Corrosion in different depths is firstly considered in simulations. The depth of the corrosion h_c varies from 1mm to 5mm whilst its radius is fixed at 2mm. The simulated signals of S_{11} and S_{21} against different corrosion depths are shown in Figure 7. It can be found that the detection probe is sensitive to the depth of corrosion. For the corrosion with variable depth, the S-parameter signals changes significantly compared with the corrosion-free case. This indicates the capability of the proposed probe in detection and evaluation of corrosion in planar conductors. It is also noticeable from Figure 7 that the variation in the signal of S_{11} is more significant than that in the S_{21} signal, which implies that the S_{11}-based evaluation of corrosion depth could have more sensitivity than that based on S_{21}. This opens up the sensitivity analysis regarding S-parameters in detection and evaluation of corrosion.

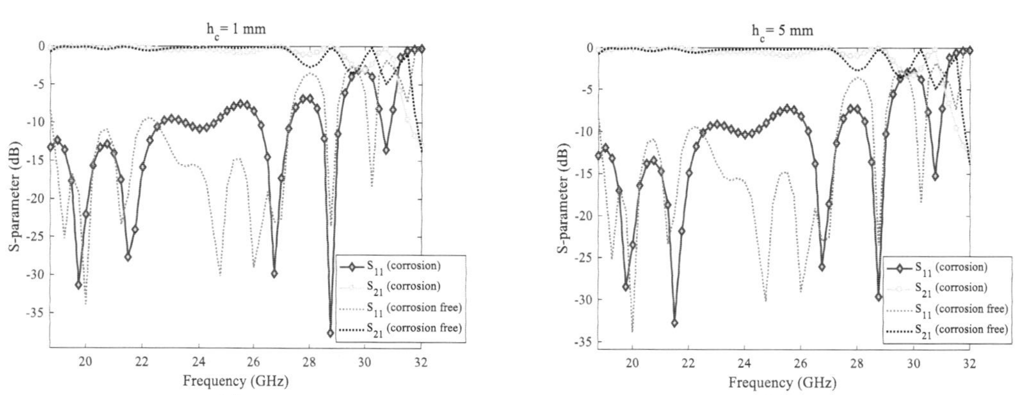

(a) (b)

Figure 7. Simulated signals of S_{11} and S_{21} for corrosion with the depth of: (a) 1mm; and (b) 5mm

In order to estimate the depth of corrosion, it is indispensable to extract signal features from the acquired signals. A new feature based on the excitation frequency and S-parameters is proposed, and calculated via:

$$F_1 = \sum_{f_o=f_{start}}^{f_o=f_{end}} (f_o \cdot S_{21}); \; F_2 = \sum_{f_o=f_{start}}^{f_o=f_{end}} (f_o \cdot S_{11}) \qquad (4)$$

where, f_{start} and f_{end} are the starting and stopping frequencies of the frequency band, respectively. f_o denotes the operating frequency whilst S_n (n=21, 11) is the corresponding S parameter at the operating frequency. The correlations of the computed F_1 and F_2 with the corrosion depth are shown in Figure 8(a). It is noteworthy that when computing F_1 and F_2, f_{start}=18.75GHz and f_{end}=32GHz. It can be seen from Figure 8(a) that that both F_1 and F_2 have monotonic relations with the corrosion depth. When the corrosion depth increases, F_1 decreases whilst F_2 rises. The reasoning behind is that as the depth of the defect increases, the standing wave remaining in the corrosion region increases. This results in the increase and drop in magnitude/power of the transmitted and reflected waves, respectively. In consequence, S_{21} and the resulting F_1 decreases whilst S_{11} and the resulting F_2 increases. The sensitivity analysis regarding F_1 and F_2 is conducted via subtraction of F_1 and F_2 in the presence of corrosion into those for corrosion-free scenario, giving the differential feature. The comparison is shown in Figure 8(b). As can be seen in Figure 8(b) that the difference in F_1 is less than that in F_2, which indicates that F_2 is more sensitive to corrosion with variable depth than F_1. This could be because the signal strength of S_{11} is higher than that of S_{21}.

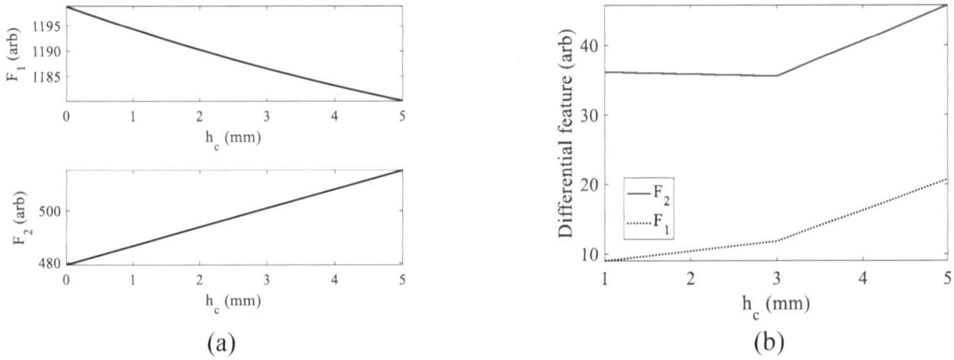

Figure 8. Analysis and comparison regarding: (a) correlations of the computed F_1 and F_2 with the corrosion depth; and (b) Differences of F_1 and F_2 in the presence of corrosion.

3.2. Corrosion in various volumes

Further investigation is focused on evaluation of corrosion with different volumes v_c. For such case, the corrosion radius varies from 1mm to 5mm whilst its depth is comparable to the radius. The simulated signals of S_{11} and S_{21} against different corrosion volumes (from 3.14mm^3 to 392.67mm^3) are shown in Figure 9. It can be seen from Figure 9 that the probe can hardly detect corrosion whose volume is 3.14mm^3. Whereas, as the corrosion volume increases, significant variation in the S-parameter signal can be found. As a result, the corrosion volume could be quantitatively evaluated based on F_1 and F_2.

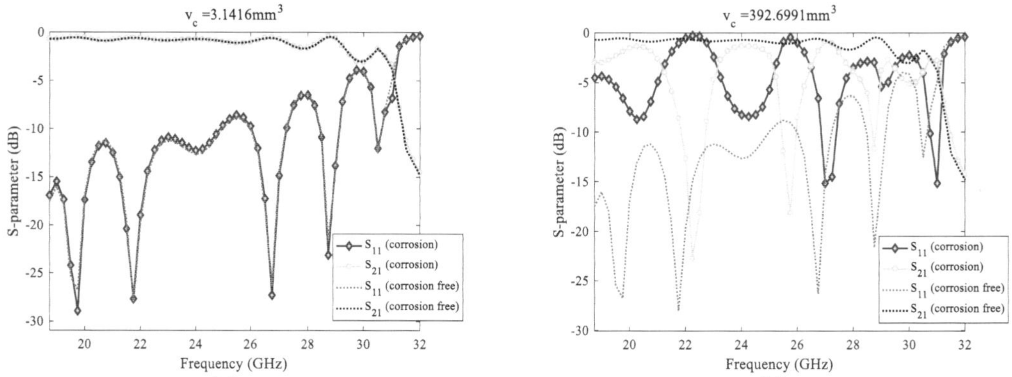

(a) (b)

Figure 9. Simulated signals of S_{11} and S_{21} for corrosion with the volume of: (a) $3.14mm^3$; and (b) $392.5mm^3$

The correlation of the computed F_1 and F_2 with the corrosion volume are presented in Figure 10(a). It can be found from Figure 10(a) that both F_1 and F_2 have monotonic relations with the corrosion volume. When the corrosion volume rises, F_1 decreases whilst F_2 rises, which can be explained by the fact presented in Section 3.1. Figure 10(b) presents the sensitivity analysis and comparison of F_1 with F_2 for evaluation of corrosion volume. It can be found in Figure 10(b) that similar to the finding presented in Section 3.1, F_2 has higher sensitivity to corrosion with different volumes than F_1. Therefore, F_2 could be preferred for high-accuracy evaluation and imaging of corrosion in planar conductors.

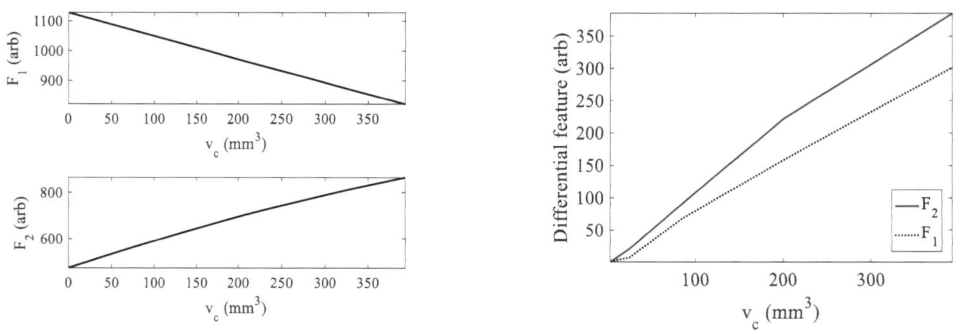

(a) (b)

Figure 10. Analysis and comparison regarding: (a) correlations of the computed F_1 and F_2 with the corrosion volume; and (b) Differences of F_1 and F_2 in the presence of corrosion.

4. Conclusion

In this paper, a new *K*-band probe has been proposed for fast screening and evaluation of corrosion in planar conductors. A 3D FE model of the proposed probe is established. Based on the model, a series of simulations are carried out for analysis of the mechanism regarding corrosion detection and characteristics of the microwave propagation within the probe cavity. The structures of 2 critical parts of the probe, i.e. the matching port and metal ring are subsequently optimised. Following this, the correlation of the testing signal of the S-parameter with corrosion in different depths and radii is intensively investigated via simulations. A new feature based on the excitation frequency and S-parameter has been proposed for corrosion evaluation. It has been found from the simulation results that the features extracted from the testing signals have monotonic relations with the depth and radius of corrosion. This would benefit the comprehensive evaluation and imaging of corrosion in planar conductors.

IOP Conf. Series: Materials Science and Engineering **554** (2019) 012003 doi:10.1088/1757-899X/554/1/012003

Acknowledgments
The authors would like to thank the Natural Science Foundation of China (Grant No. 51777149, 51477127), National Key R&D Program of China (Grant No. 2017YFF0209703), and Fundamental Research Funds for the Central Universities of China (Grant No. XJJ2018027).

References
[1] Deng Y and Liu X 2011 Electromagnetic imaging methods for nondestructive evaluation applications *Sensors* **11** 11774-808
[2] Castellano A, Foti P, Fraddosio A, Galietti U, Marzano S and Piccioni M 2015 Characterization of Material Damage by Ultrasonic Immersion Test *Procedia Engineering* **109** 395-402
[3] Zhang H, Gao B, Tian G, Woo W and Bai L 2013 Metal defects sizing and detection under thick coating using microwave NDT *NDT&E International* **60** 52-61
[4] Liu L 2010 Remote Detection and Quantitative Evaluation of Wall Thinning Volumes in a Metal Pipe *EJAM* **2** 101-109
[5] Ziehm C, Hantscher S, Hinken J, Ziep C and Richter M 2016 Near field focusing for nondestructive microwave testing at 24 GHz – Theory and experimental verification *Case Studies in Nondestructive Testing and Evaluation* **6** 70-78
[6] Yang Y, He C and Wu B 2013 Non-destructive microwave evaluation of plasma sprayed TBCs porosity *NDT&E International* **59** 34-39
[7] Liu L, Ju Y, Chen M and Fang D 2011 Application of Microwaves for Nondestructive and High-Efficiency Detection of Wall Thinning Locations in a Long-Distance Metal Pipe *Materials Transactions* **52** 2091-97
[8] Liu L, Ju Y and Chen M 2013 Optimizing the frequency range of microwaves for high-resolution evaluation of wall thinning locations in a long-distance metal pipe *NDT&E International* **57** 52-57
[9] Liu L and Ju Y 2011 A high-efficiency nondestructive method for remote detection and quantitative evaluation of pipe wall thinning using microwaves *NDT&E International* **44** 106-110
[10] Yang S, Kim K and Kang J 2013 Detection of surface crack in film-coated metals using an open-ended coaxial line sensor and dual microwave frequencies *NDT&E International* **54** 91-95
[11] Sakai Y, Yusa N and Hashizume H 2012 Nondestructive evaluation of wall thinning inside a pipe using the reflection of microwaves with the aid of signal processing *NONDESTRUCT TEST EVAL* **27** 171-184
[12] Ida N 1992 *Microwave NDT* (SPRING SCIENCE+BUSINESS MEDIA, B.V.)

IOP Conf. Series: Materials Science and Engineering **554** (2019) 012004 doi:10.1088/1757-899X/554/1/012004

Corrosion Assessment using Advanced Ultrasonic Measurement Technique

Jeffry Jamil[1], Syed Yusainee Syed Yahya[2]

[1] Advantech Alliance Sdn. Bhd, No. 12, Jalan Angkasawan U1/39, Shah Alam, Selangor, Malaysia
[2] Faculty of Applied Sciences, Universiti Teknologi MARA, Shah Alam, Selangor, Malaysia

Corresponding author, email: jeffry.jamil@advantech.my

Abstract. This paper describes the remaining wall thickness assessment using Ultrasonic Testing Phased Array (UTPA) technique. A steel plate of thickness 15 mm consists of machined simulated corrosion with various thicknesses, depths, shapes, and sizes were used in this work. Manual corrosion scanner is attached with GE Phased Array (PA) thirty-two elements (transmitter and receiver) Dual Matrix probe of 5 MHz with 1.5 mm pitch is used. The scanner then applied firmly on the plate to execute the inspection. From the result, the relevant indications from the A, B and C-Scans display are collected and analysed. The outcome of the study revealed that UTPA technique is capable to collect a high density of measuring point of thickness compared to the conventional UT technique such as Ultrasonic Testing Thickness Gauge (UTTG).

1. Introduction

Corrosion assessment is essential to estimate the service life of the pipeline, tank, boiler, pressure vessels etc. by assessing the extent of damage due to erosion or corrosion. The reduction in wall thickness of engineering structure will cause catastrophic asset failures and forced shutdown in plant. Conventional ultrasonic testing such as UTTG is extensively used to measure the remaining thickness or defect sizing of engineering structure. However, the current practice facing several limitations such as time consumption, low probability of detection (POD) and dependent on operator experiences [1].

Alternatively, the limitations of conventional UT method can be resolved by advanced UT technique. Advanced UTPA is a multi-element of small crystals housed in a single unit. The phased array beam can be steered and shaped electronically. For example, each element can be pulsed individually and varying by the time pulses along the row of elements. Desired angle and focus can be achieved by implying the constructive interference principle of the individual waves [2].

The current conventional UT crystal probe using in conventional UT has limited movement and angle of beam divergence produce high probability to miss defects whereas UTPA can optimizing the POD of defects due to sweeping the propagation of waves as shown in Figure 1.

IOP Conf. Series: Materials Science and Engineering **554** (2019) 012004 doi:10.1088/1757-899X/554/1/012004

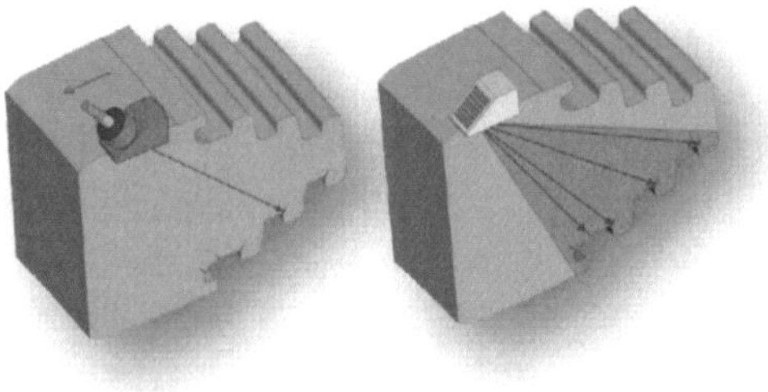

Figure 1. Comparison between capability of conventional UT and UTPA.

2. Results and discussion

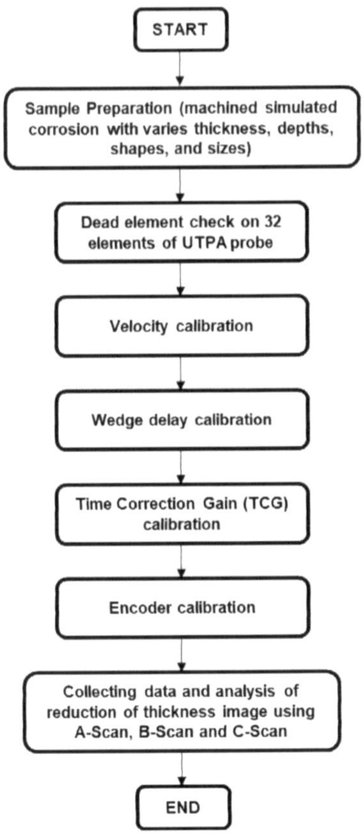

Figure 2. The flow chart of corrosion assessment using UTPA technique

The flow chart as shown in Figure 2 shows the study begin with sample preparation using machined simulated corrosion sample with various thicknesses, depths, shapes of corrosions and sizes as shown in Figure 3 and Table 1, 2 and 3 respectively below.

IOP Conf. Series: Materials Science and Engineering **554** (2019) 012004 doi:10.1088/1757-899X/554/1/012004

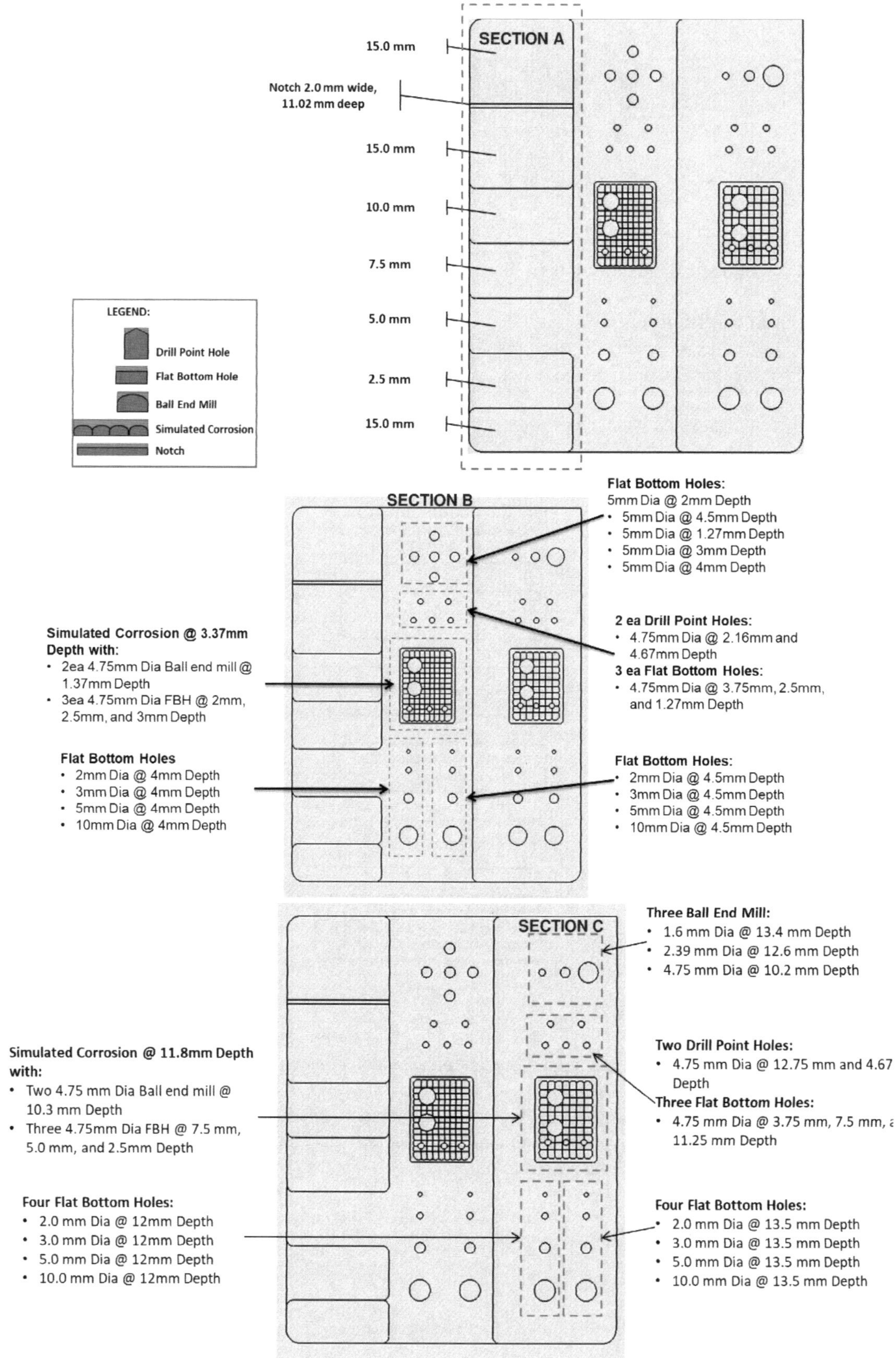

IOP Conf. Series: Materials Science and Engineering **554** (2019) 012004 doi:10.1088/1757-899X/554/1/012004

Figure 3. Specific information on machined simulated corrosion divided with section A, B and C.

Table 1. Varies thickness of fabricated simulated corrosion, Section A

Section A (all unit in mm)	
Thickness 1	15.0
Thickness 2	2.5
Thickness 3	5.0
Thickness 4	7.5
Thickness 5	10.0
Thickness 6	15.0
Notch, 2.0 mm wide	
Thickness 7	11.02

Table 2. Varies thickness of fabricated simulated corrosion, Section B

Section B (all unit in mm)	
Five flat bottom holes with 5.0 mm diameter	
Depth 1	2.0
Depth 2	4.5
Depth 3	1.27
Depth 4	3.0
Depth 5	4.0
Two drill point holes with 4.75 mm diameter	
Depth 6	2.16
Depth 7	4.67
Three flat bottom holes with 4.75 mm diameter	
Depth 8	3.75
Depth 9	2.5
Depth 10	1.27
Simulated corrosion with 3.37 mm depth	
Two ball end mill with 4.75 mm diameter	
Depth 11	1.37
Depth 12	1.37
Three flat bottom holes with 4.75 mm diameter	
Depth 13	2.0
Depth 14	2.5
Depth 15	3.0
Four flat bottom holes with deferent diameters	
Diameter	**Depth**
2.0	4.0
3.0	
5.0	
10.0	
Four flat bottom holes with deferent diameters	
Diameter	**Depth**
2.0	4.75
3.0	
5.0	
10.0	

Table 3. Varies thickness of fabricated simulated corrosion, Section C

Section C (all unit in mm)		
	Three ball end mill with different diameter and depth	
	Diameter	**Depth**
	1.6	13.4
	2.39	12.6
	4.75	10.2
	Two drill point holes with 4.75 mm diameter	
Depth 1	12.75	
Depth 2	4.67	
	Three flat bottom holes with 4.75 mm diameter	
Depth 3	3.75	
Depth 4	7.5	
Depth 5	11.25	
	Simulated corrosion with 11.8 mm depth	
	Two ball end mill with 4.75 mm diameter	
Depth 6	10.3	
Depth 7	10.3	
	Three flat bottom holes with 4.75 mm diameter	
Depth 8	7.5	
Depth 9	5.0	
Depth 10	2.5	
	Four flat bottom holes with deferent diameters	
	Diameter	**Depth**
	2.0	
	3.0	12.0
	5.0	
	10.0	
	Four flat bottom holes with deferent diameters	
	Diameter	**Depth**
	2.0	
	3.0	13.5
	5.0	
	10.0	

The testing was carried out using MENTOR UT and the apps or software was developed using MENTOR CRREAT. The dead element check was performed using GE's Phased Array (PA) Dual Multi (DM) probe of 5 MHz with 1.5 mm pitch to ensure all the thirty-two elements (transmitter and receiver) within the probe is in good condition. The velocity of calibration block was determined by attaching the probe firmly on two different thickness of step-block, which is 5.0 mm and 15.0 mm to acquire the velocity (5960 m/s) of carbon steel in the system. The determination of delay of sound propagation between the probe and the exit point of the focal law in the wedge by applying the probe firmly at 15.0 mm thickness of step-block. Time Correction Gain (TCG) calibration was calibrated and recorded with on thickness 2.5 mm, 5.0 mm, 7.5 mm, 10.0 mm, and 15.0 mm. This a method is to compensating for a reduction in signal amplitude with increasing range from thickness or reflectors of equal area. This is achieved by increasing the system gain with time so that the signals appear of equal amplitude at 80% Full-Screen-Height (FSH). The encoder for x and y-axis on scanner need to be calibrated to ensure the distance recorded in the system is within the acceptance tolerance. Lastly, data of remaining thickness were collected and analysed using A, B and C-scans display. All the methodology are shown in Figure 4.

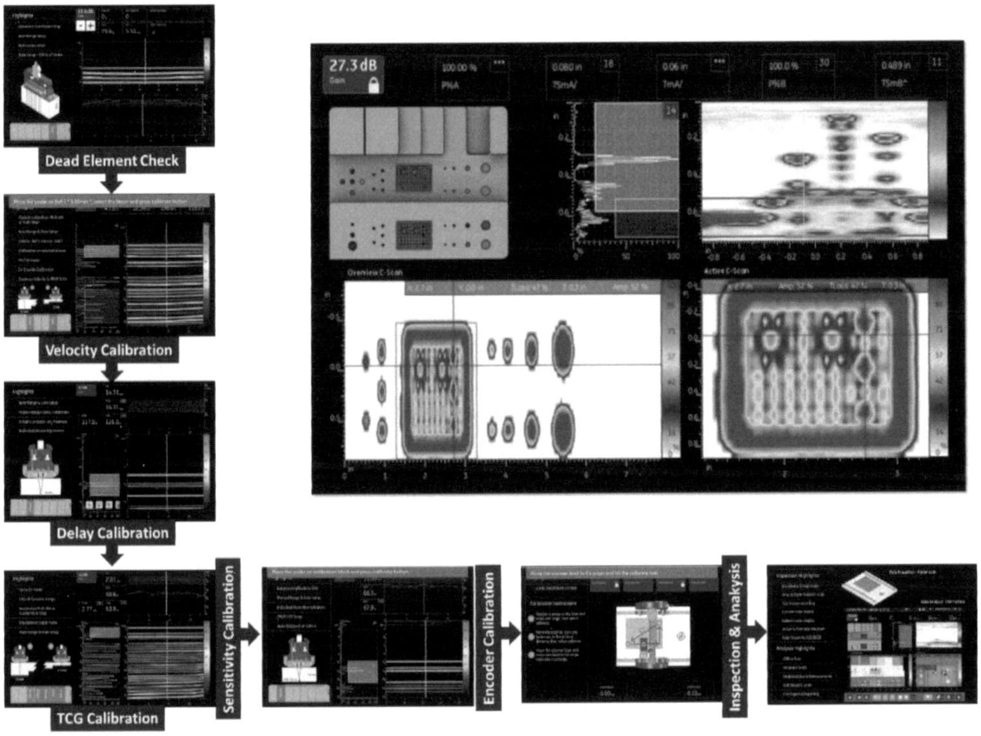

Figure 4. Procedure of UTPA calibration, inspection and remaining thickness analysis

The projected C-scan image or corrosion mapping from the UTPA is converted to CSV format as shown in Figure. 5. The result revealed a high-density of measurement point given of grid size of 1 mm × 1 mm using UTPA comparing to single point assessment capability using UTTG. Detection of true T-min generated by corrosion mapping is possible and applicable to determine and calculate the remaining life assessment of engineering structure.

4	15.04	15.04	15.03	15.03	15.03	15.03	15.04	15.04	15.03	15.04	15.04	15.06	15.06	15.05
4	15.04	15.04	15.03	12.18	12.11	12.08	12.06	12.06	12.07	12.11	12.18	15.05	15.06	15.05
4	15.04	15.04	12.11	12.05	12.01	12.01	12.01	12.01	12.01	12.03	12.06	12.11	12.22	15.05
6	15.05	12.13	12.04	12.01	12	12.01	12.01	12.01	12.01	12.03	12.03	12.06	12.12	12.25
6	15.06	12.09	12.02	12.01	12.01	12.02	12.03	12.03	12.03	12.04	12.03	12.05	12.09	12.19
6	12.21	12.07	12.01	12.01	12.01	12.03	12.03	12.03	12.03	12.04	12.03	12.05	12.08	12.16
6	12.25	12.09	12.03	12.01	12.01	12.02	12.03	12.03	12.03	12.04	12.04	12.05	12.09	12.2
6	15.06	12.16	12.06	12.03	12.01	12.02	12.02	12.03	12.03	12.04	12.05	12.08	12.16	15.07
7	15.06	15.06	12.17	12.09	12.05	12.04	12.03	12.04	12.05	12.07	12.11	12.17	15.07	15.07
7	15.07	15.06	15.06	15.06	12.19	12.14	12.13	12.13	12.14	12.19	12.3	15.07	15.07	15.07
7	15.07	15.07	15.07	15.07	15.07	15.07	15.07	15.07	15.07	15.08	15.07	15.08	15.08	15.08

Figure 5. Converted C-scan display into CSV format

Projection of 3D image as shown in Figure. 6 was used to enhance the corrosion mapping from the C-scan display to analyse the profile and condition of internal corrosion of tested product. The result also reveal the minimum thickness of the sample is 1.27 mm.

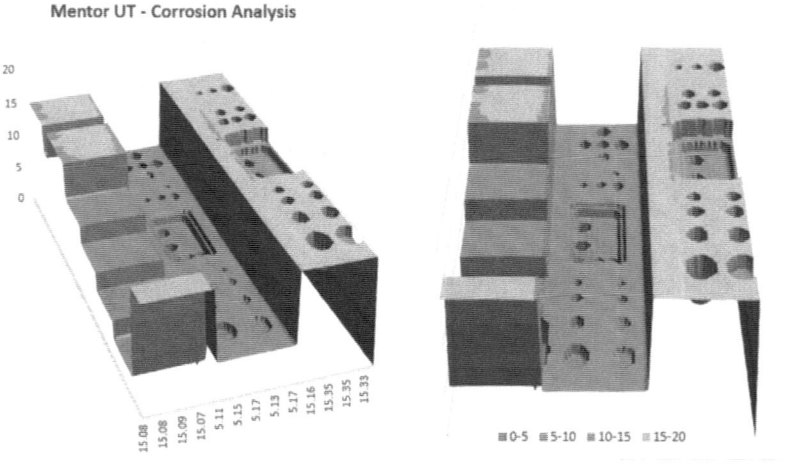

Figure 6. Projection of 3D images to identify the corrosion profile and thickness.

3. Conclusion

UTPA is an advanced technique to improve the POD of defect to avoid catastrophic disaster due to erosion and corrosion take places within the engineering product. Projected multiple displays such as A, B, and C-scans improves the detecting, locating and characterizing of defects which is essential for remaining life assessment of engineering product.

Acknowledgments

The research was supported by GE Inspection Technologies, Advantech Alliance Sdn Bhd and Faculty of Applied Sciences, University MARA of Technology (UiTM) Shah Alam.

References

[1] Le Ber L *et al.* 2016 *Corrosion Detection and Measurement Improvement using Advanced Ultrasonic Tools* 19th World Conference on Non-Destructive Testing 2016 p 1-8.
[2] Moles D C 2007 *Advances in Phased Array Ultrasonic Technology Applications* Olympus NDT p 6-9.

Feasibility of Using Radioactive Particle Tracking as an Alternative Technique for Experimental Investigation in Bubble Column Reactor

Mohd Amirul Syafiq Mohd Yunos[1], **Susan Maria Sipaun**[1], **Siti Aslina Hussain**[2]

[1] Plant Assessment Technology, Malaysian Nuclear Agency, 43000 Kajang, Selangor.
[2] Faculty of Engineering, Universiti Putra Malaysia, 43400 UPM Serdang, Selangor.

Corresponding author, email: syafiq@nuclearmalaysia.gov.my

Abstract. Radioactive particle tracking (RPT) is one of the non-invasive techniques for monitoring and investigating multiphase flow system. This technique have been widely utilized in the field of chemical process engineering for better understanding and optimizing process hydrodynamics especially in the multiphase reactor such as bubble column reactor. Due to opaque nature of industrial process systems, especially in the case of multiphase flows, non-invasive methods based on ionizing radiation have been considered for evaluating the hydrodynamic parameters. The feasibility study of radioactive particle tracking techniques in quadrilateral bubble column reactor has been successfully achieved. The radioactive particle tracking facility and data acquisition system has been developed and experimental calibration using single particle radioactive particle ^{46}Sc to investigate dynamics behaviour of quadrilateral bubble column reactor is completed. The results indicated that there is back mixing behaviour in the bubble column process. The results also reported that the radioactive particle ^{46}Sc is still in good condition and there is no radiation contamination problem arises while performing radioactive particle tracking techniques. The RPT technique was performed to reveal the instantaneous velocity and time-averaged liquid velocity in the current bubble column reactor.

1. Introduction

Before the advent of the first Malaysian pool-type nuclear research reactor TRIGA MARK II at the Malaysia Nuclear Agency (formerly known as PUSPATI) in 1982, limited access to radioactive materials, limited source quantity and high-cost of radioactive materials have restricted Malaysian researchers in their efforts for establishing nuclear applications in oil and gas industries. Large-scale utilization of radioisotopes as research and industrial investigating tools began in the late 1990s when the Plant Assessment Technology Unit, Industrial Technology Division was established under Research and Development program of Malaysian Nuclear Agency. Since then, a radioisotope that emits γ-ray has been successfully used as nuclear gauges in research and industry. The versatility of nuclear gauges innovation offers better plant troubleshooting and problem-solving ability in oil and gas industry. Recent developments in the field of radioisotope application in industry have led to a renewed interest in studying process behaviors using radiotracer techniques. Industrial radiotracers emerged as powerful tools for troubleshooting and process optimization in multiphase flow systems. However, its performance is limited by the hydrodynamic parameters and its behaviors in complex multiphase reactors. Many critical and important information failed to be extracted using a conventional method. Thus, an alternative investigation technique is required to overcome this situation.

Due to opaque nature of industrial process systems, especially in the case of multiphase flows, non-invasive methods based on ionizing radiation have been considered for evaluating the hydrodynamic parameters of multiphase reactors. These parameters include flow regimes, phase holdup, phase distributions, velocity profiles, mixing patterns and turbulence fluctuations. Thus, the demand for reliable experimental techniques to provide non-invasive, spatially, and temporally resolved data of single-phase and multiphase systems is required to assess the accuracy of complex models. The role of industrial radiotracer technology in industries has received increased attention across some disciplines in recent years. In the last decades, a significant effort to utilize γ-ray emission techniques for measuring positions and velocities in three dimensions in multiphase system reactors has been highlighted. One of the non-invasive techniques for monitoring multiphase flow system is known as a radioactive particle tracking (RPT) technique. This technique has been implemented extensively to investigate hydrodynamics characteristic in a great variety of multiphase flow systems [1].

Radioactive particle tracking (RPT) techniques has been widely utilized in the field of chemical process engineering for getting better understanding and optimizing process hydrodynamics especially in the multiphase reactor non-invasively. This technique applied a simple approach where an array of scintillation detectors is used to detect emitted isotropic γ-rays signal from radioactive particles over a sufficiently long period of time as to track the motion of that particle at a specific position and location. The radioactive particle position at certain time can be reconstructed from the widely used rigorous phenomenological model which correlates the number of photons detected and effectively measured by a scintillation detector and the positions of the emitting γ-ray radioactive particle. This strategy also involves the reconstruction of the tracer location histories known as calibration map from its real position coordinates by solving a minimization problem between the experimental and calibrated data. This is done based on tracking the motion of a neutrally buoyant single radioactive particle in order to visualize and validate the flow inside a specific volume [1,2,3,4].

There are relatively few historical studies in this technique with the first detailed study of radioactive particle tracking conducted by Kondukov *et al.* [5]. However, there is an inadequacy in his data acquisition system where quantitative results were failed to be secured. In his work, Kondukov *et al.* [5] traced a radioactive particle inside gas-solid phases in fluidized bed reactor by using 6 units scintillation detectors. Similar numbers of detectors were also used by Meek [6] for tracing solid material in turbulent liquids using different experimental design with detectors mounted on an axially moving carriage to move along with the solid particle. Unfortunately, the carriage failed to maintain its speed with tracer particle leading to data losses. Another researcher, Lin *et al.* [2] introduced first innovative computer-aided radioactive particle tracking known as CARPT with low-resolution detection from 12 unit of detectors analogue processing to investigate the solid dynamics inside the gas-solid phase of fluidized bed reactor. From then on, system transformation to the advanced radioactive particle tracking system becomes the center of attention for better hardware improvement such as those related to the data acquisition systems. The data acquisition system developed by Lin *et al.* [2] has been upgraded by Moslemian [7] where the digital pulse counts meter were introduced and has been successful in providing high-speed counting rates.

In 1991, Devanathan [3] utilized the radioactive particle tracking for investigating the first liquid phase dynamic in a gas-liquid column reactor test rig by introducing a neutrally buoyant radioactive particle [3,8]. Subsequently, the same principles have been applied by Yang, [9] to perform better numerical calculation using a weighted least-squares method in particle tracking post-processing techniques, extensively. The position rendering method proposed by Yang, [9] promoted another invention on the rigorous phenomenological relation generated using Monte Carlo simulation by Larachi *et al.* [10] followed by the development of enhanced algorithm with direct back-propagation neural network model [11].

Much of the literature since the mid-1990s emphasizes the use of non-invasive radioactive particle tracking techniques for the characterization of multiphase reactors. Research into radioactive particle tracing applications in multiphase flow system has a long history. There are a large number of published studies that describe the uses of radioactive particle tracking technique for gas–solid–liquid phase investigations in fluidized bed reactor [1,12,13,14,15,16,17,18,19,20,21,22,23,24], solid phase in cylindrical tumbler reactor [25,26], gas–solid and gas–liquid phase in risers [27,28,29,30,31], liquid phase in circulating draft tube photobioreactor [32], solid–solid phase in V-blender reactor [33], and pebble bed reactors [34,35]. However, there is very little published research on the gas-liquid phase in quadrilateral bubble column reactors which need to be comprehensively investigated and examined using radioactive particle tracking technique.

Surprisingly, there have been no preliminary works reported for the application of radioactive particle tracking techniques in Malaysia. The primary challenge faced by researchers in Malaysian is the limited capability to produce or purchase radioactive material at a reasonable cost. In addition, the radioactive particle tracking hardware and portable data acquisition system is currently not suitable for field testing in the petrochemical industry. This scenario indicates the need to set up innovative RPT facility for hydrodynamics investigation of industrial-scale reactors in Malaysia. Therefore, this study seeks to remedy those situations by designing and utilizing radioactive particle tracking as an alternative technique for the first experimental investigation in bubble column reactor in Malaysia. Furthermore, this study also set out to gain further understanding of the feasibility of using this alternative method for gas-liquid phase behavior. A specialized radioactive particle which neutrally buoyant in water phase will be used together with the fixed radiation detectors around the quadrilateral bubble column reactor. This radioactive particle tracking technique is using the simulated calibration map using MCNPX code to track the particle position in order to reconstruct the particle trajectory and visualize the velocimetry behavior of the water phase in bubble column reactor. The development of data acquisition system hardware and software will also be discussed.

2. Methodology

2.1. Experimental Setup

The experiments were carried out in a quadrilateral bubble column reactor test rig with an internal diameter of 200 mm and a height of 1800 mm. Figure 1 shows a schematic diagram of a quadrilateral bubble column reactor used for radioactive particle tracking measurement. The quadrilateral column was fabricated using Perspex® acrylic glass (density = 1.18 g/cm^3) with inner cross-section equal to 200 mm x 200 mm and wall thickness equal to 3 mm. The column is filled up with 0.04 m^3 normal tap water of ambient pressure and at a room temperature. The height of the fluid volume (water level) referred to the stationary liquid level was fixed at 1000 mm. The air as a gas phase was supplied from centralized compressors with a working pressure of 1.5 MPa. A pressure regulator and flow meters regulated the air flow rate at a fixed rate of 20 liters per minutes. Air is purged into the reactor to create air bubbles for aeration from the bottom by passing through perforated air distributor known as sparger plate at the bottom of the bubble column reactor. The sparger design D used in this study has a porosity of 0.18% and the total opening cross-sectional area of 70.86 mm^2. It consists of 361 holes of 0.5 mm diameter with the triangular pitch of 1 cm apart.

Radioactive particle tracking facility consists of an array of radiation detectors and data acquisition systems. Ten (10) units 2 inch x 2 inch (5.08 cm diameter x 5.08 cm long) crystal of sodium iodide activated with thallium, NaI (Tl) scintillation detectors (crystal density = 3.67 g/cm^3) are strategically fixed around the bubble column reactor as can be seen in Figure 1. The detectors are positioned on five planes alternately, and the distance between the consecutive planes is equal to 20 cm. The detectors are held by aluminium extrusion profile bar structures which in turn are supported on stainless steel round bars with dimension 3.81 cm x 3.81 cm positioned at 90° intervals around the column. Two detectors are placed at one level to face each other, surrounding the column with 180° spacing. The distance between the surface of the detector and the column wall is fixed at 5 cm equally. The arrangement of the detectors in each plane is shown in Figure 2.

When radiation passed each scintillation detector, the radiation counts were collected instantaneously and recorded into a computer. The hardware components used for this experiment is a multi-channel

IOP Conf. Series: Materials Science and Engineering **554** (2019) 012005 doi:10.1088/1757-899X/554/1/012005

analyzer Ludlum Model 4612 Counter, and a 12-detector single channel analyzer (SCA) with RS232 control to compute the signals from the scintillation detectors. Block diagram of the data acquisition system is shown in Figure 3. The desktop computer or industrial durable notebook is used to control all the parameters, obtain and display data, and extract results. The physical picture of radioactive particle tracking setup for bubble column reactor can be seen in Figure 4. This versatile data acquisition hardware system set up is safe to be used in oil and gas industries. The compact set of the multi-channel analyzer, scintillation detector and durable notebook are using minimal electrical power source. Therefore, the set up can reduce the potential to induce spark and ignition that increase the risk of fire and explosion inside refinery plants.

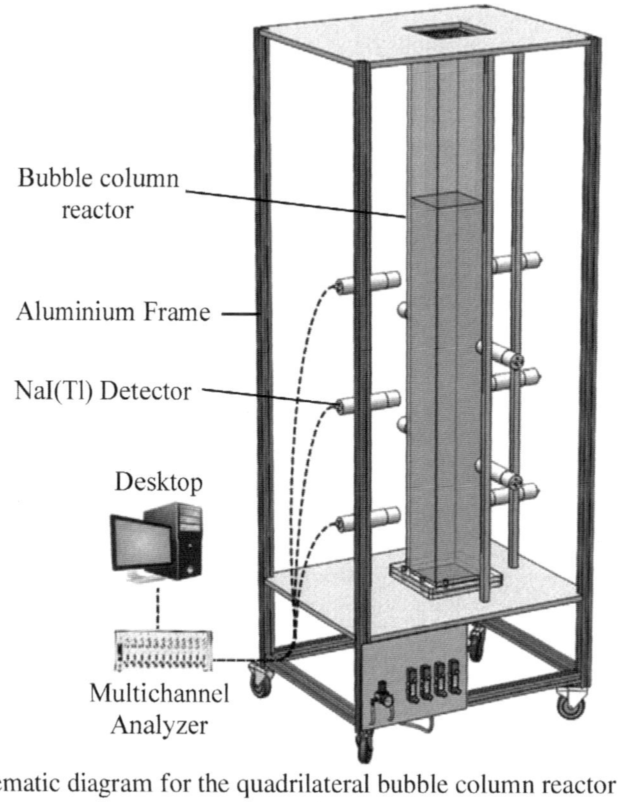

Figure 1. Schematic diagram for the quadrilateral bubble column reactor and radioactive particle tracking setup.

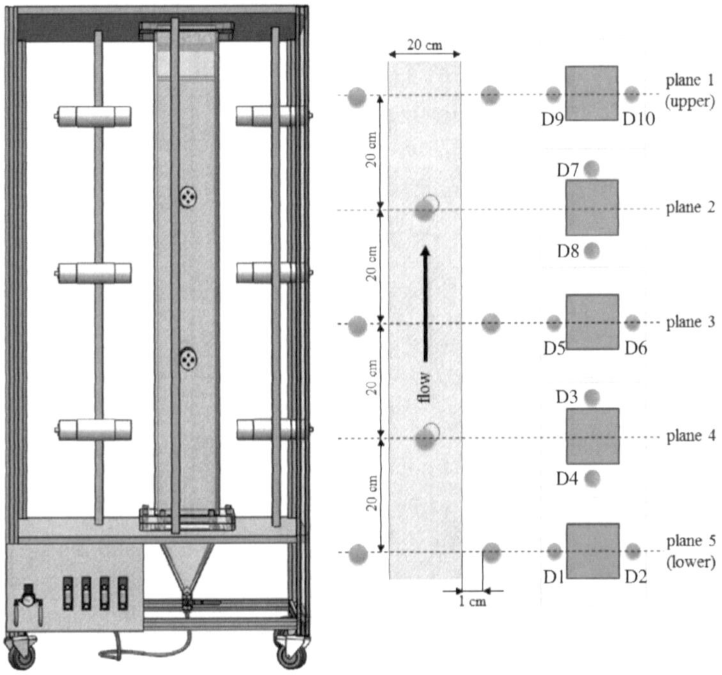

Figure 2. The arrangement of the scintillation detectors in experimental setup

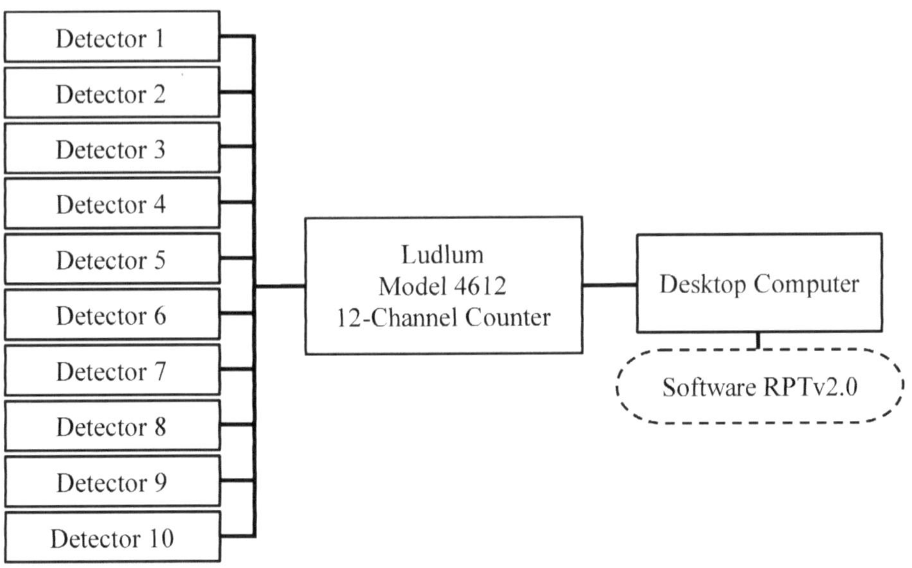

Figure 3. Block diagram of the data acquisition system.

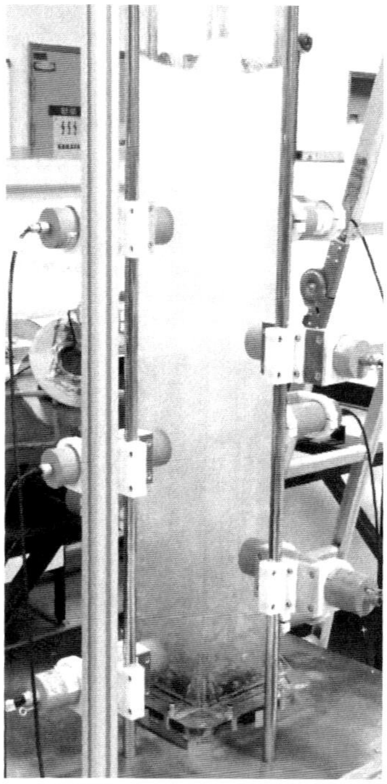

a. RPT array of scintillation detector

b. MCA Ludlum Model 4612

c. Data acquisition system

Figure 4. Physical pictures of RPT experimental setup.

2.2. The Radioactive Particle

Due to various advantages, a scandium radioactive particle with an activity of 4 mCi was selected for this experiment. This selection is made based on the fact that the radioisotope ^{46}Sc has nuclear transitions emitting gamma radiation with sufficient energy for a radioactive particle tracking application. Scandium has a half-life of 83.8 days and presents two photopeaks, one at 889 keV and one at 1121 keV. In radioactive particle tracking technique, a single tiny radioactive particle that is naturally buoyant is used for tracking the motion of liquid phase in bubble column reactor. Enough space is required to maintain the buoyancy of the particle equivalent to water density. Taking this fact into consideration, a hole is made in polyethylene bead that contains the scandium glass and the empty gap of the hole is filled sufficiently with the Araldite epoxy glue to prevent radioisotope dilution, dispersion, and leakage that may lead to radioactive contamination inside and outside the reactor. The air gap and amount of the glue inside the particle are adjusted to match the density of water. About 20 mg scandium glass (Sc_2O_3) was encapsulated in polypropylene beads sphere particles of diameter 2–3 mm so that the effective density of the tracer particle is closely matched with the density of liquid phase. Final target density practically close to the water density of 1.001 kg/m^3 at ambient temperature where the neutrally buoyant particle is achieved. The γ-ray emitter ^{46}Sc radioactive particles are activated by thermal neutron activation in a nuclear reactor. In this study, the particle was irradiated at 750 kW for 6 hours with neutron flux density of 1.2×10^{12} thermal neutrons per cm^2/s inside the core of TRIGA-Mark II reactor at the Malaysian Nuclear Agency.

2.3. RPT Data Acquisition Software

Detection of single radioactive which emits gamma radiation was achieved by the use of a scintillation detector without collimator positioned around the bubble column. The response signals produced by the detector were recorded continuously at the specific interval period of time. Data acquisition companion software for acquiring data and recording response signals from sets of scintillation detector in radioactive particle tracking experiment initially developed by Rahman *et al.* [36] for study the hydrodynamic behaviour in quadrilateral bubble column reactor. National Instruments LabView 8.6 system design software has been utilized to develop radioactive particle tracking data acquisition system software. This system software consists of an interactive scanning control graphical user interface (GUI), a data acquisition from 12-channel RS-232 serial interface counter, and a multichannel analyser (MCA). Rahman *et al.* [36] reported that the second version of the software was successfully upgraded and currently known as RPTv2.0. The graphical user interface (GUI) for this software includes three operating modes that consist of system setup and configuration mode, scanning mode and multichannel analyser, upper-level discriminator, and lower-level discriminator (MCA & LLD/ULD) setup mode.

The system setup and configuration mode enabled the configuration of COMM port for communication between the multichannel analyzer and the desktop host communication port. This can be seen in Figure 5. Operating system input setup such as total scanning time, the configuration setting, and output files setup are available in this mode. Before running the data acquisition mode or scan mode, a suitable window of the region of interest for gamma radiation spectrum must be set in the software. Thus, the MCA and LLD/ULD setup mode shown in Figure 6 can be used to configure the lower and upper levels of gamma spectrum with communication between the software and a multichannel analyzer. This mode is designed to help the user in selecting the appropriate values for LLD and ULD parameters between 0 to 3300 millivolts. The graphical user interface of LLD/ULD adjustment of gamma spectrum result for ^{46}Sc radioactive particle tracer is shown in Figure 7. The graphical user interface for scanning mode is shown in Figure 5. In this mode, the integration time (counts rate) and the total period of acquisition time both have to be determined before starting the data acquisition process. The integration time describes the duration of each data count and period of acquisition means the total time of the data acquisition process. Ludlum Model 4612 multichannel analyser has the capability to obtain the data counts at -higher speeds up to 50 ms per count. The scan mode also allows user to control the start and stop the scanning process. The updated count signals are displayed with respect to the integration time and the scanning process is continued until desired counting time is reached. Immediately upon the completion of data acquisition process, a pop-up graphical user interface will inform the user to save the acquired results in the specified folder location [36]. Figure 8 shows the flowchart for data acquisition system process by utilizing RPT hardware and software.

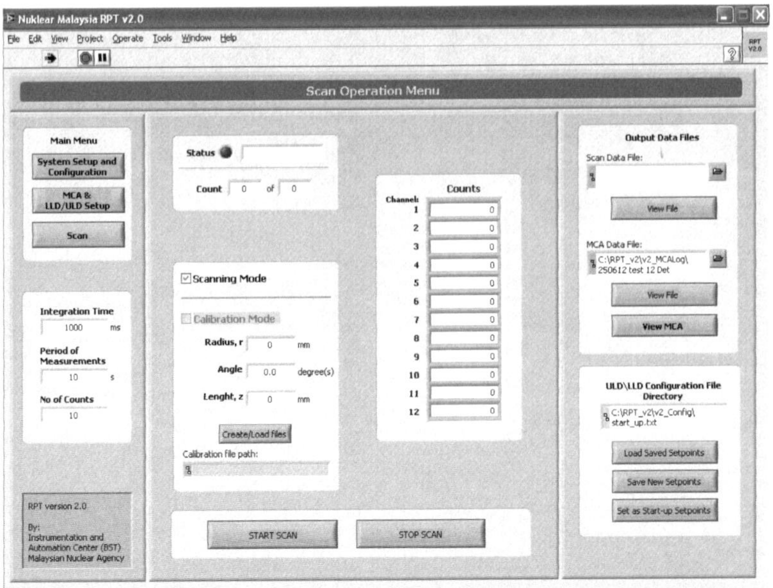

Figure 5. The graphical user interface (GUI) for RPT software version 2.0.

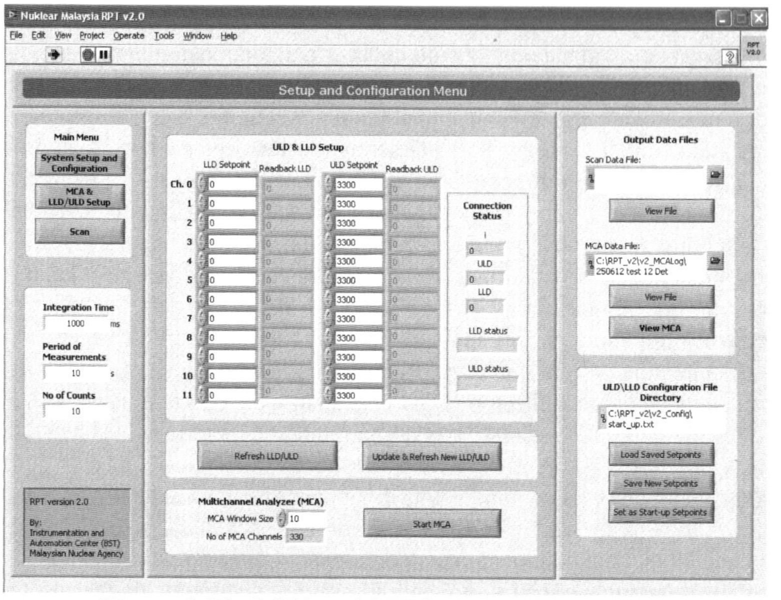

Figure 6. The MCA and LLD/ULD setup mode GUI.

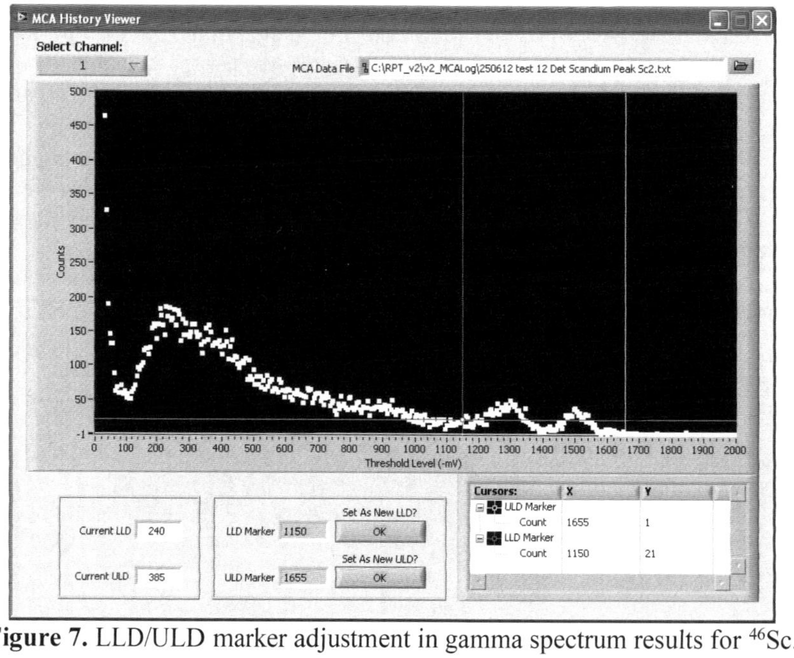

Figure 7. LLD/ULD marker adjustment in gamma spectrum results for ^{46}Sc.

IOP Conf. Series: Materials Science and Engineering **554** (2019) 012005 doi:10.1088/1757-899X/554/1/012005

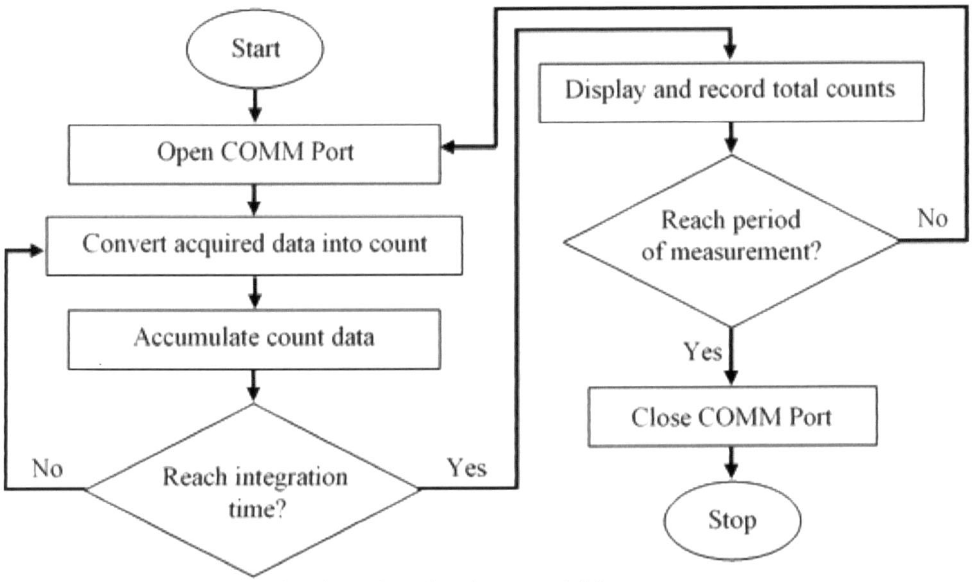

Figure 8. Flow chart for data acquisition system process.

3. Results and Discussions

The present study was designed to determine the feasibility of using radioactive particle tracking as an alternative technique for hydrodynamics behavior investigation in bubble column reactor. The radioactive particle tracking facility has been successfully developed. The facility includes the system hardware that consists of pilot-scale quadrilateral bubble column reactor, ten sets of scintillation detector with special detector holder, and data acquisition system that used compact and versatile multichannel analyzer. One of the primary concern in using alternative radiotracer method is the radiological safety to radiation workers and members of public. In this study, encapsulated ^{46}Sc radioactive particle inside polypropylene bead was successfully prepared to prevent contamination inside the reactor vessel.

Before the RPT experiment was carried out, the position of the particle has to be reconstructed accurately from the analyzed radiation signals. The particles will be placed at known positions depending on the size and volume of the vessels to be mapped. There are few methods used to calibrate the particle position in radioactive particle tracking technique. First method uses a calibration device which was able to calibrate the locations of radioactive particle in angular, axial and radial directions using the automatic computerized device. This device is known as computer-aided radioactive particle tracking (CARPT) [13]. The second method is using position reconstruction by utilizing the Monte Carlo simulation code for solving a minimization problem between measured counts and a rigorous phenomenal model that relates the position of the emitter to the number of counts recorded by each detector surrounding the system. The detectors measure different levels of radiation depending on the position of the emitter. Using an appropriate model, the exact location of the tracer can then be obtained.. Generally, in Monte-Carlo based reconstruction, the intensities of γ-rays emitted by tracer particle at specified locations are documented by positioning scintillation detectors around the vessel during the calibration stage.

Figure 9. Manual radioactive particle calibration holder devices.

In this study, manual calibration process using special equipment is chosen because automatic calibrations devices are currently not available for radioactive particle tracking. In order to track particle inside the bubble column reactor accurately, a conventional benchmark experiment was carried out using manual calibration equipment to validate the particle position reconstructed by Monte Carlo simulation. This equipment was fabricated using stainless steel with single radioactive particle holder as can be seen in Figure 9. The calibration equipment was designed to reconstruct particle histories. Figure 10 indicates that distance versus radiation signals relationships known as calibration curve for the detector provides the basis to reconstruct the tracer particle instantaneous position. A comparison between the simulated particle positions reconstructed using Monte Carlo algorithm and the particle positions obtained experimentally is shown in Figure 10. The comparison shows a great matching between these two sets of coordinates. MCNPX code was applied to calculate a map of counts containing a total of 2600 grid points as a function of particle coordinates. In total, 26000 data sets have been successfully obtained with ten detectors were used for the reconstruction of instantaneous positions with the aid of the 2600 known coordinates.

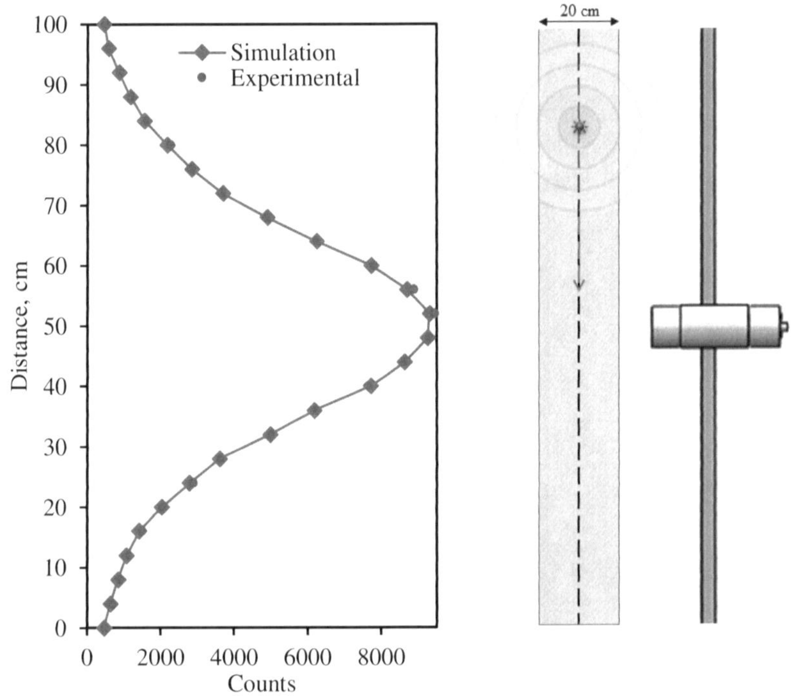

Figure 10. A comparison between MNCPX simulation and experimental distance counts rate measurement for Detector 5.

Upon completing the calibration step, the neutrally buoyant radioactive tracer particle will be left freely moving with the liquid phase inside bubble column reactor while the scintillation detectors collect count intensities at a suitable frequency (10-20 Hz) depending on the flow environments and the activity of the radioactive particle. A sampling frequency of 20 Hz and total acquisition time of 2.5 hours is used in this study. This is because the homogeneous regime conditions in the bubble column reactor with superficial gas velocity is 0.0083 m/s. Figure 11 shows the data acquisition results for ten detectors at different planes and radioactive particle ^{46}Sc axial trajectories for the acquisition time 2.5 minutes was illustrated at different levels. By using the calibration curve and the reconstruction algorithms, the Lagrangian trajectory of the radioactive tracer can be reconstructed and the result is as shown in Figure 12 (a) and (b). The experiments will be operated for a sufficient time to allow the radioactive particles numerously explore the whole region of bubble column reactor so the time-averaged velocity in each compartment reaches a plateau. This will ensure enough statistics related to radioactive tracer particle trajectories, which represents the liquid dynamics of the bubble column reactor.

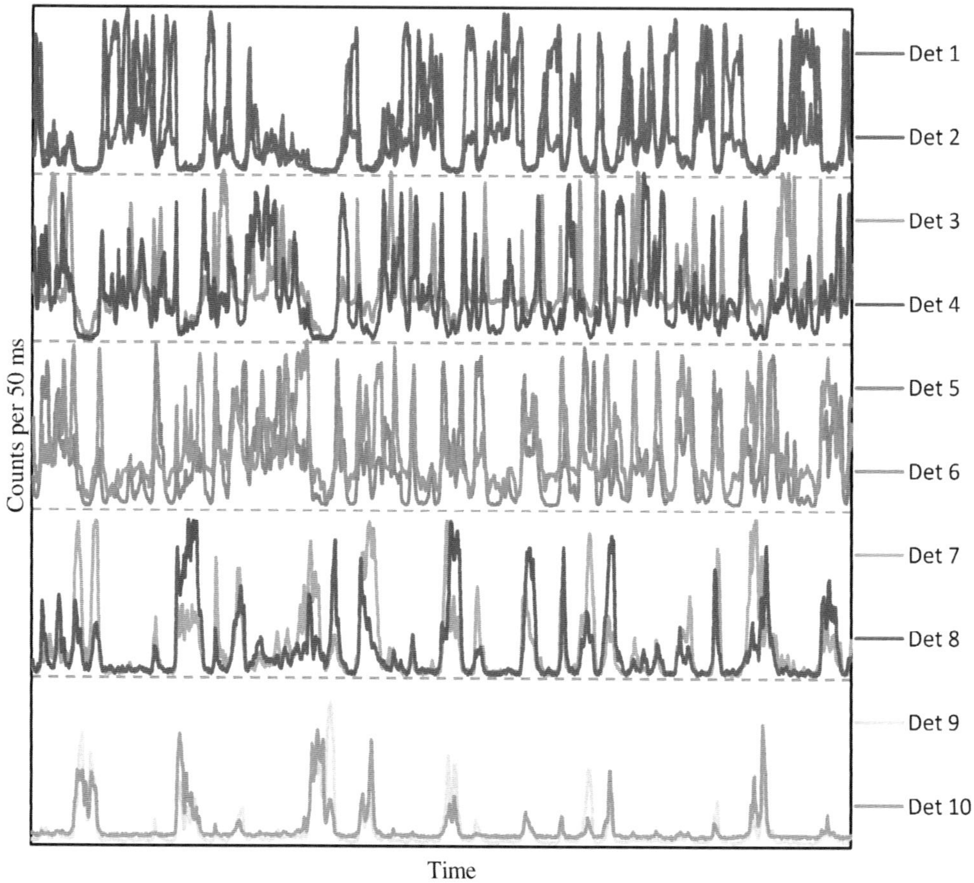

Figure 11. Radioactive particle ^{46}Sc axial trajectories for the 2.5 minutes at different detector level.

Once the distance of the particles from the set of detectors is evaluated, a weighted regression scheme is used to estimate the position of the particle at a given sampling instant in time. Thereby a set of instantaneous position data is obtained that gives the positions of the particle at successive sampling instants. Time differentiation of the successive particle positions yields the instantaneous Lagrangian velocities of the particle as a function of time and position of the particle. From the Lagrangian particle velocities, ensemble averaging is performed to calculate the average velocities of the liquid. For Cartesian coordinate system, the instantaneous velocity (u_l) and mean velocity (u) were calculated by determining the magnitude of the velocity vector (speed) as follow:

$$u_x = (x_2 - x_1)(t_2 - t_1) \tag{1}$$

$$u_y = (y_2 - y_1)/(t_2 - t_1) \tag{2}$$

$$u_z = (z_2 - z_1)/(t_2 - t_1) \tag{3}$$

$$u_1 = \{(x_2 - x_1)(t_2 - t_1)\}i + \left\{\frac{(y_2 - y_1)}{(t_2 - t_1)}\right\}j + \left\{\frac{(z_2 - z_1)}{(t_2 - t_1)}\right\}k \tag{4}$$

$$u = \sqrt{{u_x}^2 + {u_y}^2 + {u_z}^2} \tag{5}$$

IOP Conf. Series: Materials Science and Engineering **554** (2019) 012005 doi:10.1088/1757-899X/554/1/012005

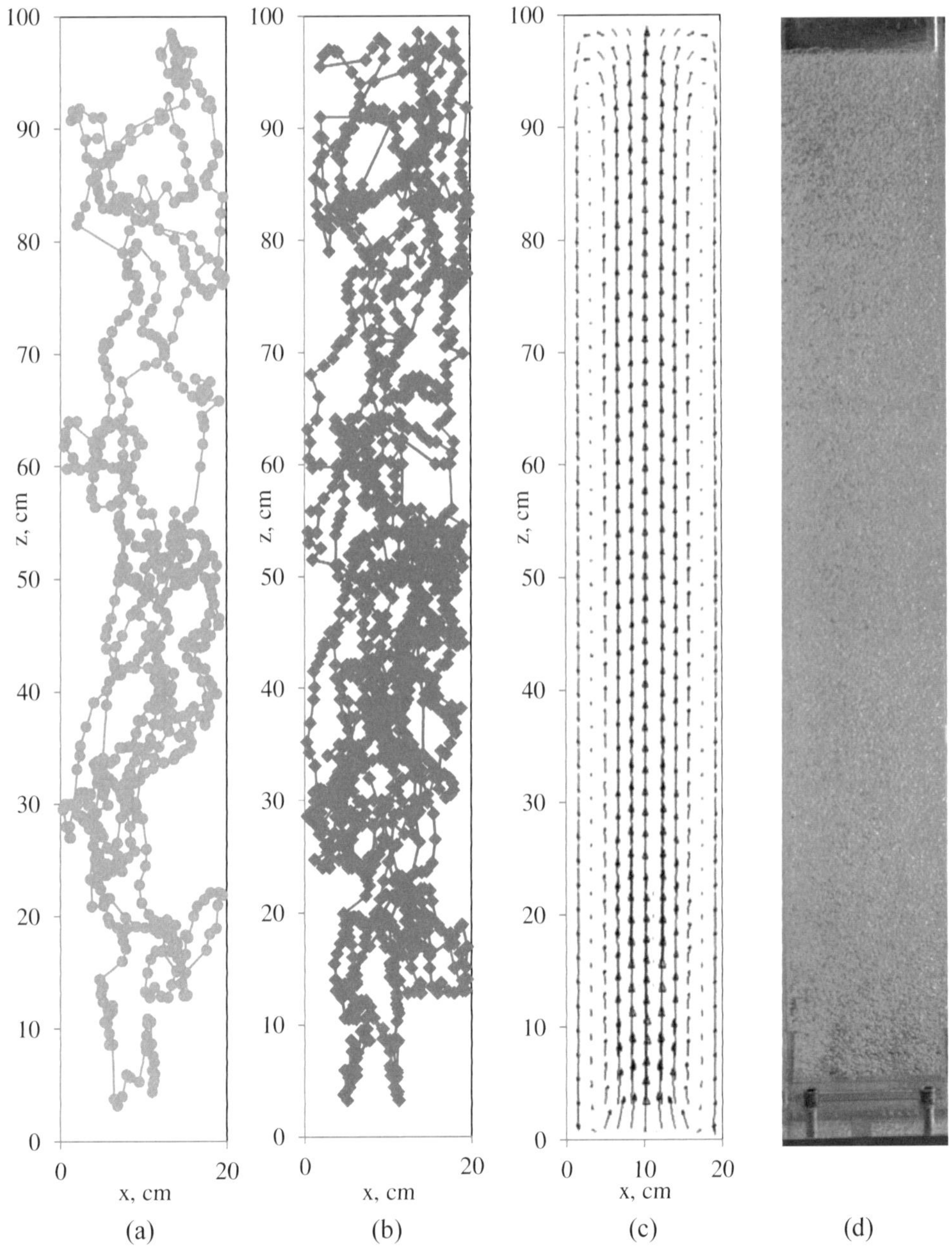

Figure 12. Lagrangian trajectory of radioactive particle ^{46}Sc in bubble column reactor with superficial gas velocity 0.083 m/s for (a) 25 sec acquisition time (b) 75 sec acquisition time. (c) Time-averaged velocity vector plots. (d) A graphical image of the dispersed bubble for sparger design type D.

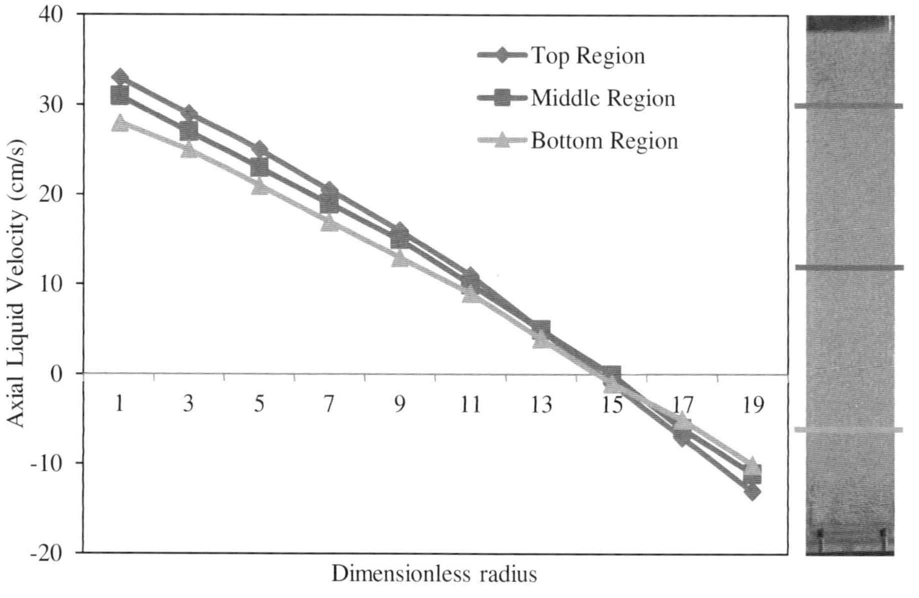

Figure 13. Comparison of axial liquid velocity at different region of column.

The instantaneous liquid velocity in Figure 12(c) has been calculated by time differentiating between two successive locations of the tracer particle. The time-averaged liquid velocity is obtained by averaging the instantaneous velocity of each cell by taking the mean of the number of occurrences obtained for each sampling compartment as shown in Figure 13. The results presented the comparison between bubble column region at top, middle and bottom of the vessel. Thus, this study concluded that the alternative RPT technique is also capable of providing information about the fluid dynamic parameters.

4. Conclusions

The feasibility study of radioactive particle tracking techniques has been successfully completed. The radioactive particle tracking facility has been developed and experimental calibration using single particle radioactive particle ^{46}Sc to investigate dynamics behavior of quadrilateral bubble column reactor has been accomplished. The RPT technique was performed to reveal the instantaneous velocity and time-averaged liquid velocity of bubble column reactor. The results indicated that there is back mixing behavior in the bubble column process. The results also reported that the radioactive particle ^{46}Sc is still intact and there is no radiation contamination problem arises while performing radioactive particle tracking techniques.

Acknowledgements

The authors gratefully acknowledge financial support by the Ministry of Science, Technology and Innovation Malaysia in the frame of Science Fund Grant (03-03-01-SF0245) and International Atomic Energy Agency (IAEA) through Technical Cooperation Project Number RAS/1/012 and Research Contract No.17374. The authors are also gratefully acknowledged staffs of the Faculty of Engineering, Universiti Putra Malaysia, Industrial Technology Division, and Malaysian Nuclear Agency for their support and supervision. Thanks to the Nuclear Power Division for the provision of neutron irradiation and research facilities. A special acknowledgement is also attributed to members of Plant Assessment Technology Group (PAT), who have involved in the design, fabrication, experimental setup and facilities for enabling this work to be presented.

IOP Conf. Series: Materials Science and Engineering **554** (2019) 012005 doi:10.1088/1757-899X/554/1/012005

References

[1] Bhusarapu S, Al-Dahhan MH, Dudukovic MP (2006) Solids flow mapping in a gas-solid riser: Mean holdup and velocity fields *Powder Technology* 163 **1–2** 98–123
[2] Lin JS, Chen MM, Chao BT (1985) A novel radioactive particle tracking facility for measuring solids motion in gas-fluidized beds *AIChE Journal* **31** 465–473
[3] Devanathan N, Moslemian D, Dudukovic MP (1990) Flow mapping in bubble columns using CARPT *Chemical Engineering Science* 45 **8** 2285–2291
[4] Roy S, Larachi F, Al-Dahhan MH, Dudukovic MP (2002) Optimal design of radioactive particle tracking experiments for flow mapping in opaque multiphase reactors *Applied Radiation and Isotopes* **56** 485–503
[5] Kondukov NB, Kornilaev AN, Skachko IM, Akhromenkov AA, Kruglov AS (1964) An investigation of the parameters of moving particles in a fluidized bed by a radioisotopic method *Int. Chem. Eng* 4 **1** 43
[6] Meek CC (1972) Statistical characterization of dilute particulate suspensions in turbulent fluid fields *Ph.D. Thesis*, University of Illinois, Urbana, Illinois
[7] Moslemian D, Chen MM, Chao BT (1987) Influence of solids hydrodynamics on local heat transfer from tube banks immersed in a gas fluidized bed *American Society of Mechanical Engineers*, New York, NY
[8] Azizi S, Yadav A, Lau YM, Hampel U, Roy S, Schubert M (2017) On the experimental investigation of gas-liquid flow in bubble columns using ultrafast X-ray tomography and radioactive particle tracking *Chemical Engineering Science* **170** 320–331
[9] Yang YB, Devanathan N, Dudukovic MP (1992) Liquid back mixing in bubble columns via computer automated particle tracking (CARPT) *Chemical Engineering Science* **47** 2859–2864
[10] Larachi F, Kennedy G, Chaouki J (1994) A γ-ray detection system for 3-D particle tracking in multiphase reactors *Nuclear Instruments and Methods in Physics Research Section A: Accelerators, Spectrometers, Detectors and Associated Equipment* 338 **2–3** 568–576
[11] Godfroy L, Larachi F, Kennedy G, Grandjean B, Chaouki J (1997) Online flow visualization in multiphase reactors using neural networks *Applied Radiation and Isotopes* 48 **2** 225–235
[12] Cassanello M, Larachi F, Marie MN, Guy C, Chaouki J (1995) Experimental characterization of the solid phase chaotic dynamics in three-phase fluidization *Industrial & Engineering Chemistry Research* 34 **9** 2971–2980
[13] Larachi F, Cassanello M, Chaouki J, Guy C (1996) Flow structure of the solids in a 3-D gas–liquid–solid fluidized bed *AIChE Journal* 42 **9** 2439–2452
[14] Kiared K, Larachi F, Guy C, Chaouki J (1997) Trajectory length and residence-time distributions of the solids in three-phase fluidized beds *Chemical Engineering Science* 52 **21–22** 3931–3939
[15] Kiared K, Larachi F, Chaouki J, Guy C (1999) Mean and turbulent particle velocity in the fully developed region of a three-phase fluidised bed *Chem. Eng. Technol.* **22** 683–689
[16] Fraguío MS, Cassanello MC, Larachi F, Chaouki J (2006) Flow regime transition pointers in three-phase fluidized beds inferred from a solid tracer trajectory *Chemical Engineering and Processing* **45** 350–358
[17] Fraguio MS, Cassanello MC, Larachi F, Limtrakul S, Dudukovic M (2007) Classifying flow regimes in three-phase fluidized beds from CARPT experiments *Chemical Engineering Science* **62** 7523–7529
[18] Mabrouk R, Chaouki J, Guy C (2007) Effective drag coefficient investigation in the acceleration zone of an upward gas–solid flow *Chem. Eng. Sci.* 62 **1–2** 318–327
[19] Lefebvre S, Guy C, Chaouki J (2007) A convective/dispersive solid phase mixing model for three-phase fluidized bed reactors: Effect of dimensionless numbers *Chem. Eng. Sci.* 62 **18–20** 4954–4962
[20] Ahuja GN, & Patwardhan AW (2008) CFD and experimental studies of solids hold-up distribution and circulation patterns in gas–solid fluidized beds *Chemical Engineering Journal* 143 **1–3** 147–160

[21] Khanna P (2007) Particle tracking in a lab-scale conical fluidized bed dryer (*Doctoral dissertation*)

[22] Upadhyay RK, Roy S (2010) Investigation of hydrodynamics of binary fluidized beds via radioactive particle tracking and dual source densitometry *The Canadian Journal of Chemical Engineering* **88** 601–610

[23] Tamadondar MR, Azizpour H, Zarghami R, Mostoufi N, Chaouki J (2012) Using particle trajectory for determining the fluidization regime in gas–solid fluidized beds *Adv. Powder Technology* 23 **3** 349–351

[24] Jain S, Saraswat P, Jain V, Pant HJ, Upadhyay RK (2014) Investigation of liquid–solids fluidized bed of different particle size through radioactive particle tracking techniques *Journal of Radioanalytical and Nuclear Chemistry* 302 **3** 1309–1313

[25] Sheritt RG, Chaouki J, Mehrotra AK, Behie LA (2003) Axial dispersion in the three-dimensional mixing of particles in a rotating drum reactor *Chem. Eng. Sci.* 58 **2** 401–415

[26] Alizadeh E, Dubé O, Bertrand F, Chaouki J (2013) Characterization of mixing and size segregation in a rotating drum by a particle tracking method *AIChE J.* 59 **6** 1894–1905

[27] Roy S, Chen J, Kumar SB, Al-Dahhan MH, Duduković MP (1997) Tomographic and particle tracking studies in a liquid–solid riser *Industrial & Engineering Chemistry Research* 36 **11** 4666–4669

[28] Roy S, Kemoun A, Al-Dahhan MH, Dudukovic MP (2005) Experimental investigation of the hydrodynamics in a liquid–solid riser *AIChE Journal* 51 **3** 802–835

[29] Bhusarapu S, Al-Dahhan M, Dudukovic MP (2004) Quantification of solids flow in a gas-solid riser: Single radioactive particle tracking *Chemical Engineering Science* **59** 5381–5386

[30] Mahmoudi S, Baeyens J, Seville J (2011) The solids flow in the CFB-riser quantified by single radioactive particle tracking *Powder Technology* 211 **1** 135–143

[31] Upadhyay RK, Roy S, Pant HJ (2012) Benchmarking radioactive particle tracking (RPT) with laser Doppler anemometry (LDA) *International Journal of Chemical Reactor Engineering* **10** 1

[32] Jung SH, Park JG, Moon JH, Kim JB (2015) Investigation of hydrodynamic parameters in a draft tube reactor using radioisotope based techniques and conventional method *Applied Radiation and Isotopes* **101** 93–100

[33] Doucet J, Bertrand F, Chaouki J (2008) An extended radioactive particle tracking method for systems with irregular moving boundaries *Powder Technology* 181 **2** 195–204

[34] Wang Z, Lee KO, Gardner RP (2014) A dual system for monitoring the positions of multiple radioactive tracer pebbles in scaled pebble bed reactors *Nuclear Technology*, 185 **3** 259–269

[35] Khane V, Said IA, Al-Dahhan MH (2016) Experimental investigation of pebble flow dynamics using radioactive particle tracking technique in a scaled-down Pebble Bed Modular Reactor (PBMR) *Nuclear Engineering and Design* **302** 1–11

[36] Rahman NAA, Yussup N, Abdullah JB, Ibrahim MBM, Abdullah NA, Mokhtar MB, Hassan HB (2015) Development of Labview based data acquisition and multichannel analyzer software for radioactive particle tracking system *AIP Conference Proceedings* **1659** 030012, AIP Publishing

Investigation of the effect of metallic screens on image quality in gamma computed radiography

D'Ademo, N.

DÜRR NDT GmbH & Co. KG, Höpfigheimer Straße 22, Bietigheim-Bissingen, Germany

Corresponding author, email: dademo.n@duerr-ndt.com

Abstract. The role of metallic screens in computed radiography for controlling scattered radiation is very well-known. Likewise, the intensification effects of certain lead screen thicknesses with higher X-ray energies, such as those produced by gamma isotopes is also a topic that has been recently investigated in literature. This paper extends this topic further and quantitatively investigates how image quality in terms of signal-to-noise ratio (SNR) and basic spatial resolution (SR_b) is influenced by metallic screens when Iridium-192 is used as the radiation source. The result data is then used to make recommendations to achieve optimum image quality when using Iridium-192.

1. Introduction

1.1. Previous Work

This study is largely inspired by a 2016 paper [1] by Steven Mango which investigated a similar topic and provided the following conclusions and recommendations:

(1) Lead screens have a huge influence on image intensity (pixel value) in gamma exposures, especially with higher-energy isotopes such as Iridium-192. Specifically, thin lead front screens produce a greater intensification effect (maximum intensification at 0.125 mm thickness) compared to thick lead front screens. This effect is also present with rear lead screens but conversely thicker screens give more intensification than thinner screens (maximum intensification at 0.250 mm thickness). Furthermore, copper screens do not produce this intensification effect at all.

(2) This intensification is desirable in front screens as it produces an increase in image signal-to-noise ratio (SNR). However, in the case of rear screens, it is undesirable as it could cause increased unsharpness (reduced SR_b) due to the fact that it is primarily low-energy scatter radiation incident on the rear of the imaging plate that is being amplified.

(3) Consequently, the rear lead screen should not be in direct contact with the imaging plate, but instead an additional shielding of copper (between 0.125 mm and 0.250 mm is sufficient) or steel should be placed in-between in order to absorb this undesirable fluorescence produced by the lead screen.

IOP Conf. Series: Materials Science and Engineering **554** (2019) 012006 doi:10.1088/1757-899X/554/1/012006

1.2. Aims
This paper has the following aims:
- To confirm the aforementioned main findings of [1].
- To quantitatively measure the effects of various metallic screen thicknesses and types on image quality by measuring signal-to-noise ratio (SNR) and basic spatial resolution (SR_b) as well as the associated pixel value.
- To propose practical recommendations for using metallic screens with Iridium-192 based on the result data.

1.3. Theory
Although the detailed physics involved in the interaction between X-ray radiation and matter is beyond the scope of this paper, it important to understand on a high level the reasons why screens may be used in digital radiography. This topic is explored in more detail in [1] and can serve as a useful additional introduction.

1.3.1. Scattered Radiation
When a photon with sufficiently high energy collides with an orbital electron of an atom within a given material, the electron is ejected and the photon "scatters" in a different direction and with a lower energy. These scattered photons are of interest due to the fact that imaging plates have higher absorption efficiency at lower radiation energies and therefore low-energy photons traveling in an altered direction (compared to the primary X-ray beam) may have a significant effect on image quality. Thin lead screens are typically used both on the front and back of the imaging plate to absorb these undesirable photons.

1.3.2. Fluorescence

Fluorescence also known as the photoelectric effect. In this case, the incoming photon is completely absorbed by the atom when it collides with the electron (which is then ejected and becomes a "photoelectron"). Lower energy photons are produced when an electron from a higher shell drops down to the lower shell to fill the vacancy created by the ejected electron – the energy of these photons depends on the energy level difference between the two electron shells (which is characteristic for a given element). Like before, this low-energy radiation is readily absorbed by the imaging plate. For the X-ray energy levels typically used in industrial radiography, the energy required to eject electrons from the innermost shell (i.e. most strongly bounded electrons) is most relevant – the energy at which this begins to occur is known as the "K-edge" and differs depending on the material. As we increase the incoming X-ray energy above the K-edge, the probability of this photoelectric absorption occurring decreases. Note: As the K-edge of lead is approximately 88 keV (compared with 9 keV for copper), we can expect to observe this fluorescence when using lead screens in industrial radiography.

2. Methodology
2.1. Exposure Setup
In order to accurately measure the effect various screen types and thicknesses have on image quality, it is necessary to carefully ensure that any parameter that could potentially change the image quality is kept constant while the only variable that changes is the screen setup. Figure 1 and Table 1 detail the setup used throughout the entire experiment.

The following considerations were also taken to further ensure reliable data:
- All exposures were taken in a single day (over a 9-hour period) in order to reduce the influence of reducing isotope source activity.
- The same location and equipment (imaging plate, computed radiography scanner, screens, gamma source, etc.) was used throughout the entire experiment.

- The imaging plate was scanned immediately after the exposure and was read by the scanner in same orientation/direction in every scan.
- A lower sensitivity imaging plate (HD-IP) was used in combination with a longer exposure time (as opposed to a high-sensitivity IP and short exposure time) in order to reduce the influence of any slight variations in exposure time (i.e. dosage).
- The object was placed directly on top of a 16.3 mm wood board in order to allow scatter radiation to reach the rear of the imaging plate.
- The stackup from the source-side was as follows: Pb → Cu → Object → IP → Cu → Pb

Table 1. Exposure setup

Radiation Source	
Type	Iridium-192 (with collimator)
Activity	25.19 Ci
Focal Size	3 mm
Object	
Material	Stainless Steel (SS304)
Thickness	10 mm
Setup	
Imaging Plate	10x24 cm HD-IP
Source-to-Object Distance	300 mm
Exposure Time	4:00 min
Scanner	
Type	DÜRR NDT HD-CR 35 NDT
Laser Size	25 μm
Reading Pitch	25 μm

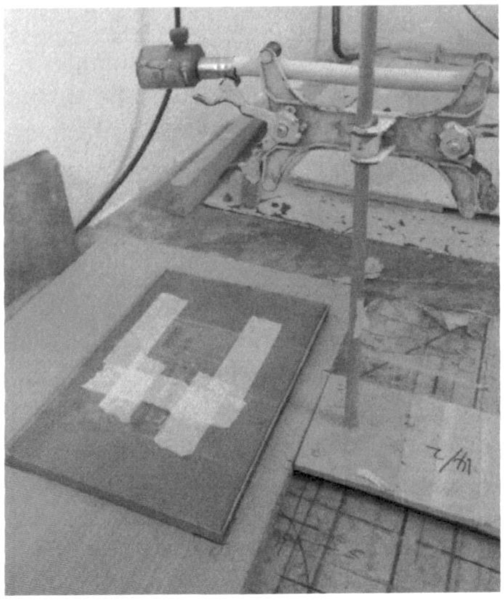

Figure 1. Exposure setup

2.2. Image Quality Measurement

IOP Conf. Series: Materials Science and Engineering **554** (2019) 012006 doi:10.1088/1757-899X/554/1/012006

For each screen setup, a single exposure was performed and the following image quality metrics were measured using the DÜRR NDT D-Tect image analysis software:

2.2.1. Signal-to-Noise Ratio (SNR)
Mean value of 5 different areas (each with dimensions of 100x100 pixels).

2.2.2. Pixel Value.
Mean value of same 5 areas used for the SNR measurements.

2.2.3. Basic Spatial Resolution (SR_b). Measured using a source-side duplex wire IQI in accordance to ISO 19232-5 [2]. The modulation (dip) for duplex wire pairs 1D to 5D was measured over the entire width of the IQI (651 pixel lines) and the average of these five values was recorded as a measure of the image basic spatial resolution (Note: The 2nd-order interpolation method was not used as it consistently yielded incorrect results when performed with only five duplex wire pairs).

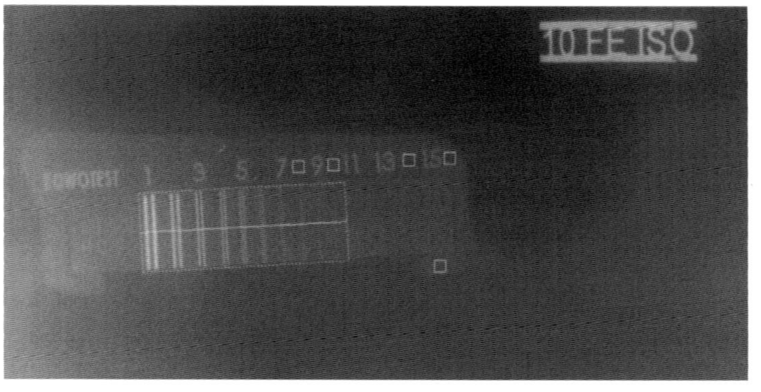

Figure 2. Cropped digital X-ray image from D-Tect software showing the areas of measurement.

3. Results

All result data is normalized to an exposure with no front and no rear screens. Furthermore, all graphs have been plotted with the same y-axis (vertical axis) range so that the relative effects of each screen configuration can be visually compared easily. For reference, the numerical value for each plotted data point can be found in the summary table (Table 2) at the end of this section.

3.1. Front Screen Only

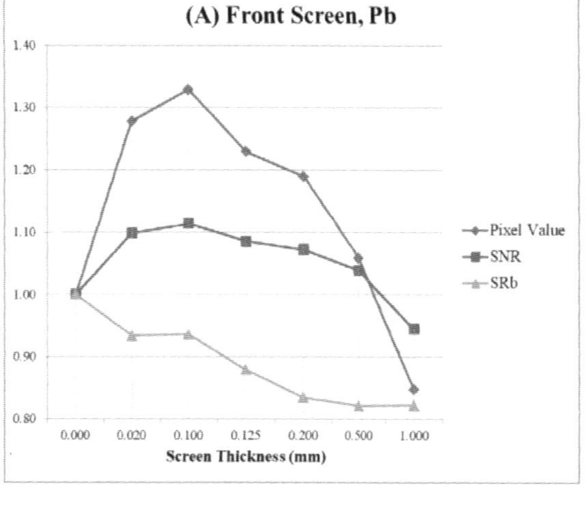

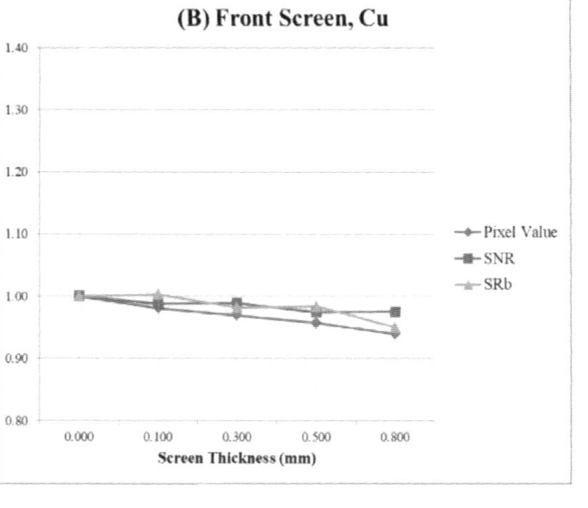

Figure 3. Effects of lead (A) and copper (B) front screen thickness on image quality.

3.1.1. *Observations*

- As expected, a higher pixel value corresponds to a higher image signal-to-noise ratio (SNR). This correlation is non-linear.
- Only lead screens provide intensification - copper front screens do not produce any intensification at all (instead they only attenuate the incoming radiation).
- Intensification due to the lead screen increased until a maximum at 0.100 mm thickness. The incoming radiation began to be attenuated at a lead screen thickness of between 0.500 mm and 1 mm.
- Basic spatial resolution (SR_b) seemed to decrease with lead screens until the intensification effect was no longer present (which can be interpolated to occur at slightly above 0.500 mm lead thickness).
- Compared to lead screens, copper screens gave significantly less degradation in basic spatial resolution (SR_b).

3.2. *Rear Screen Only*

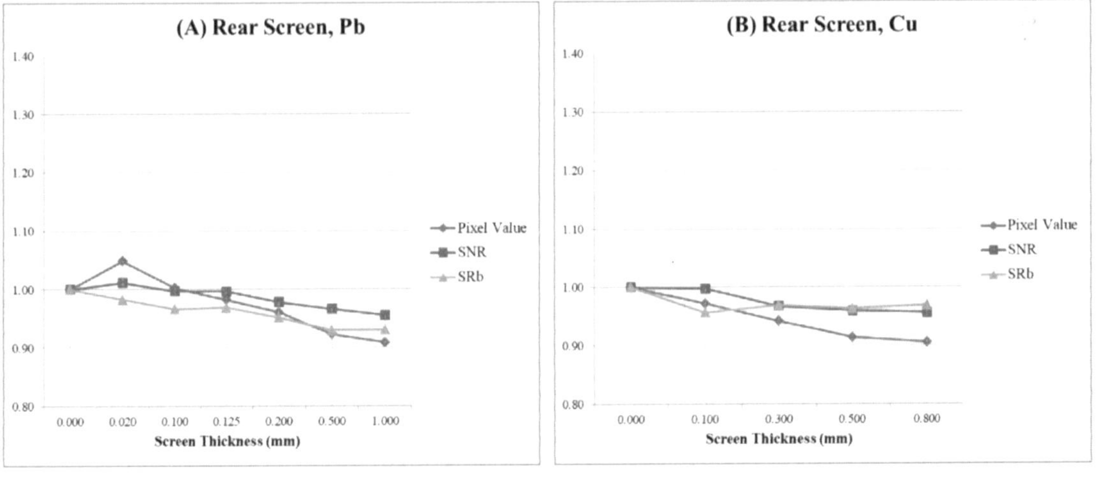

Figure 4. Effects of lead (A) and copper (B) rear screen thickness on image quality.

3.2.1. *Observations*

- As in the case with front screens, only lead rear screens provide intensification (not copper); however, the effect is dramatically less for rear screens compared to front screens. Both 0.020 mm and 0.100 mm lead screens provided marginal intensification – thicknesses above this attenuated the radiation.
- As expected, pixel value is strongly correlated to the signal-to-noise ratio (SNR).
- Both lead and copper screens did not seem to improve the basic spatial resolution (SR_b); however, the degradation was much less with copper.

3.3. Front & Rear Screen

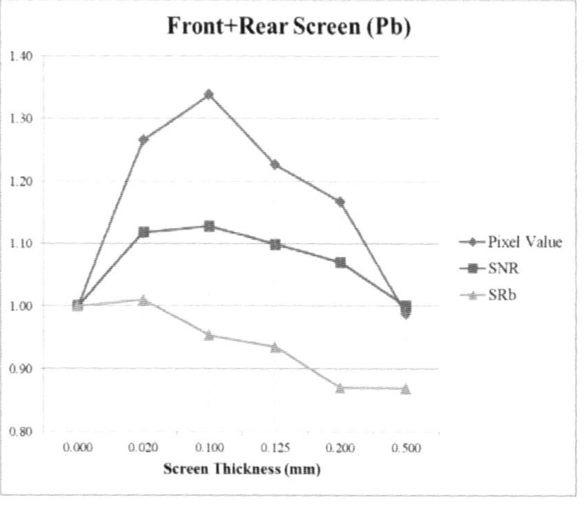

Figure 5. Effects of front and rear lead screen thickness on image quality.

3.3.1. Observations

- The overall results match closely to what we would expect if we combined together the previous results for front lead screen (Figure 3A) and rear lead screen (Figure 4A).
- Basic spatial resolution (SR_b) is better when a rear lead screen is used (as opposed to using only a front lead screen).

3.4. Practical Screen Combinations

Since a 0.100 mm lead screen was found to produce the most intensification, the effects of using this particular thickness with various copper screen thicknesses were investigated for both the front and rear of the imaging plate.

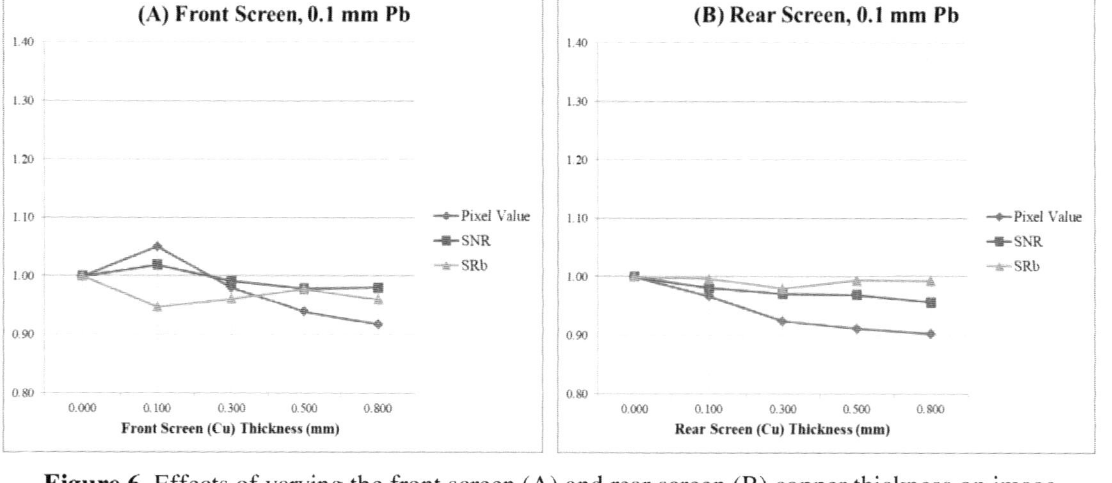

Figure 6. Effects of varying the front screen (A) and rear screen (B) copper thickness on image quality. A 0.100 mm lead screen is used in both cases. Note: The 0.000 mm Cu thickness data corresponds to the no screen exposure.

IOP Conf. Series: Materials Science and Engineering **554** (2019) 012006 doi:10.1088/1757-899X/554/1/012006

3.4.1. Observations

- In comparison to using only a lead screen, the use of a copper screen in between the imaging plate and lead screen improved the basic spatial resolution (SR_b). This was the case for both the front and rear of the imaging plate.
- All of the fluorescence produced by the 0.100 mm front lead screen is absorbed by a copper screen with thickness 0.300 mm and above.
- With a rear copper screen thickness of 0.300 mm in combination with a 0.100 mm lead screen, most of the radiation incident on the rear of the imaging plate can be attenuated (i.e. pixel value decreases only slightly for copper screen thicknesses above this).

IOP Conf. Series: Materials Science and Engineering **554** (2019) 012006 doi:10.1088/1757-899X/554/1/012006

4. Data

Table 2. Complete result dataset (Note: Empty cells correspond to 'no screen').

Figure	Front Screen (mm)		Rear Screen (mm)		Pixel Value	SNR	SR_b
	Pb	Cu	Pb	Cu			
-					1.00	1.00	1.00
3A	0.020				1.28	1.10	0.93
	0.100				1.33	1.11	0.94
	0.125				1.23	1.08	0.88
	0.200				1.19	1.07	0.83
	0.500				1.06	1.04	0.82
	1.000				0.85	0.95	0.82
3B		0.100			0.98	0.99	1.00
		0.300			0.97	0.99	0.98
		0.500			0.96	0.98	0.98
		0.800			0.94	0.98	0.95
4A			0.020		1.05	1.01	0.98
			0.100		1.00	1.00	0.97
			0.125		0.98	1.00	0.97
			0.200		0.96	0.98	0.95
			0.500		0.92	0.97	0.93
			1.000		0.91	0.96	0.93
4B				0.100	0.97	1.00	0.96
				0.300	0.94	0.97	0.97
				0.500	0.91	0.96	0.96
				0.800	0.91	0.96	0.97
5	0.020		0.020		1.27	1.12	1.01
	0.100		0.100		1.34	1.13	0.95
	0.125		0.125		1.23	1.10	0.94
	0.200		0.200		1.17	1.07	0.87
	0.500		0.500		0.99	1.00	0.87
6A	0.100	0.100			1.05	1.02	0.95
	0.100	0.300			0.98	0.99	0.96
	0.100	0.500			0.94	0.98	0.98
	0.100	0.800			0.92	0.98	0.96
6B			0.100	0.100	0.97	0.98	1.00
			0.100	0.300	0.92	0.97	0.98
			0.100	0.500	0.91	0.97	0.99
			0.100	0.800	0.90	0.96	0.99

5. Conclusions

Since the data collected in this investigation is largely dependent on the exposure setup, the following findings may not apply to every inspection scenario; nevertheless, they should serve as a useful starting point when using Iridium-192:

(1) A 0.100 mm front lead screen gives the largest intensification effect (>30% increase in pixel value resulting in a >10% increase in SNR).
(2) There is a trade-off when using a copper front screen in combination with a lead front screen:
   a. The benefit of increased SNR due to the intensifying effect (fluorescence) of the lead screen as mentioned previously in (1) is almost eliminated.
   b. However, a copper screen will give a slightly sharper image (in comparison to using only a lead screen).
(3) Using a 0.100 mm lead screen with a 0.300 mm copper screen on the rear of the imaging plate provides approximately the same amount of backscatter attenuation as a single 0.500 mm lead screen. This rear screen combination also produces a sharper image (in comparison to using only a lead screen).

Considering that it is always desirable to reduce the exposure time for radiation safety reasons (in particular when using a portable gamma isotope such as Iridium-192), the use of a 0.100 mm front lead screen is a simple way to achieve this while still allowing imaging plate flexibility if inspecting curved objects.

Moreover, the overall findings mostly match those of [1]; however, the experiment presented in this paper measured significantly lower rear lead screen intensification and furthermore, increasing the rear lead screen thickness did not produce increased intensification (but instead attenuated the radiation).

5.1. Future Work

In almost all screen setups, basic spatial resolution (SR_b) was measured to be lower than what was achieved when using no screens at all. It was not tested whether increasing the dose could have compensated for this slight reduction in sharpness due to the use of screens; thus, a further experiment could investigate whether using for example a copper front screen with a longer exposure time increases the image sharpness (and if so, to what extent).

Furthermore, any future similar study should ideally use averaged data from multiple identical exposures in order to improve the accuracy of the results (due to time constraints this was not possible in this experiment).

Acknowledgements

The author would like to thank Nuklear Malaysia, in particular Dr Khairul Anuar for allowing use of their facility and equipment in order to collect the data required to perform this study. Special thanks also to Zulfahmy Awaldin, Ahmad Khuzairi Rastam, Engku Mohammad Imran, Daniel Chow and the entire NDT Instruments team both in Malaysia and Singapore for their support throughout this research.

References

[1] Mango, S., *Practical Considerations and Effects of Metallic Screen Fluorescence and Backscatter Control in Gamma Computed Radiography*, 19th World Conference on Non-Destructive Testing, Munich, Germany, 2016
[2] *Non-destructive testing – Image quality of radiographs, Part 5: Determination of the image unsharpness and basic spatial resolution value using duplex wire-type image quality indicators*, ISO 19232-5, 3rd ed., Aug 2018

IOP Conf. Series: Materials Science and Engineering **554** (2019) 012007 doi:10.1088/1757-899X/554/1/012007

Preliminary Study of Digital Laminography Parameters on Carbon Steel Plate

Shahfuan Hanif A. Hamidi[1], Khairul Anuar M. Salleh[2], Ab Razak Hamzah[1], Mahamad Noor Wahab[1], Noorhazleena Azaman[2] and Anis Salwa Ahmad[1]

[1]Fabrication & Joining Section, Universiti Kuala Lumpur- Malaysia France Institute, 43650, Selangor, Malaysia
[2]Industrial Technology Division, Malaysian Nuclear Agency, 43600, Selangor, Malaysia

Corresponding author, email: shahfuanh@unikl.edu.my

Abstract. Laminography is a technique used to determine the depth of a defect in Radiographic Testing (RT). It enables RT operators to accurately locate the defect from the object's surface especially for welded components. The technique will ensure the quality of weld components meets the standard requirement for reliability and safety purposes. This paper intends to study the exposure parameters of laminography technique on carbon steel plate by using Radiographic Testing – Digital. (RT-D). The important exposure parameters that are considered in this preliminary study are voltage, current, and frame time. The radiography images with achievable level of contrast sensitivity and image quality are analyzed according to the ISO 17636-2. Discussion on correlation between the normalized Signal to Noise ratio (SNR_N) and the exposure parameters are done based on the plotting graphs. The results show the exposure parameters has significantly influenced the value of SNR_N. Further experimental works are performed using triangulation technique on welded sample. The deviation between the calculated depth of defects and actual results is less than 1.0mm.

1. Introduction

Identification of welding defects is important to determine the integrity and fit for purpose of a welded component in steel structures. Cracks, cavities, solid inclusion, lack of fusion/penetration, and imperfect dimension, are examples of defects which are not homogenous to the parent metal due to welding activities and may lead to failure during in-service condition [1].

Carbon steels are the most widely recognized and used in steel structures. Low carbon, medium carbon, and high carbon steel materials are highly demanded for large and heavy components. because of the lower cost [1]. Structures made of carbon steel are typically welded to form a strong and sturdy build up.

Welding is a joining process involving melting and solidification for homogenous metallurgical composition. An understanding on welding metallurgy of carbon steels is required for excellent mechanical properties, chemical composition, and weldability as well as to ensure the welded components are free from any defects [2].

Characterization of welding defects by various Non-Destructive Testing (NDT) techniques is a critical part in welding inspection of steel structures. A primary, simplest, easy to apply, fast, and relatively inexpensive technique in inspection is by visual examination [3]. However this technique has a limitation to evaluate the sub-surface defects. It requires one or more of the other NDT techniques

IOP Conf. Series: Materials Science and Engineering **554** (2019) 012007 doi:10.1088/1757-899X/554/1/012007

such as Radiographic Testing (RT), Ultrasonic Testing (UT), Penetrant Testing (PT), or Magnetic Testing (MT) for a better quality of welded steel structure that meets the requirements for safety and reliability [4].

Radiographic Testing (RT) is a sophisticated NDT technique that is commonly used to evaluate almost all carbon steels surface and sub-surface defects. RT employs X-rays or gamma rays to penetrate an object and indicate any defects on viewing medium; i.e. photographic film or electronic detector [1]. Over the years, RT is based on film radiography systems [5]. It is traditionally demand for an experienced operator to interpret the film radiography (RT-F) image. Manual interpretation is time consuming and sometimes due to lack experience of interpreter, the results will be inconsistent even when high resolution radiography images is provided [4]. Hence, a development of Radiographic Testing – Digital (RT-D) should be taking place as new digital technologies to overcome this issues [6].

RT-D with Digital Detector Array (DDA) system offered a lot of benefit compared to RT-F [7]. RT-D with calibrated DDA is completely tends to eliminate the noise sources arise in radiography. Essential parameters of RT-D such as basic spatial resolution of the image (SR_b^{image}) and normalized Signal-to-Noise Ratio (SNR_N) are the key measurements for high contrast, sensitivity, and image quality for visibilities of defects [8]. Hence, the optimum exposure parameters of X-ray source and DDA, i.e voltage, current, source to object distance (SOD), and frame time are required in determining achievable value of SR_b^{image} and SNR_N.

The evaluation of contrast sensitivity of radiography image used a single wire Image Quality Indicator (IQI) which contains a series of straight wires (with different diameters) and similar material with object to evaluate [9]. Following standard ISO 19232-5, a standardized duplex wire IQI which consists of 13 wires pairs of platinum with different diameters is used to determine the SR_b^{image} for requirement of image unsharpness [10]. Then, the value of SNR_N can be measured from SR_b^{image} value as stated in (1).

In this research paper, ISO 17636-2 is followed as standard practice in determining minimum requirement for image quality values of IQI and SNR_N for application of DDAs for weld inspection [11][12].

$$SNR_N = SNR_{measured} \frac{88.6\mu m}{SRb} \tag{1}$$

For determination on minimum source to object distance (SOD), f_{min} as exposure parameter, the following equation is followed [12]:

$$\frac{f_{min}}{d} \geq 7.5b^{\frac{2}{3}} \tag{2}$$

Where b is the object to detector distance while d is the focal spot size.

Early development of RT-D in carbon steel structures, it is used as a high image quality forming for visibility of defects. Evaluation depth of defects is impossible by this technology without experienced in projection techniques [13]. Laminography technique is one of the projection techniques that enables the operator to accurately evaluate the depth of a defect.

In figure 1(a), the principle of laminography is referred to the relative motion of the X-ray source, the detector and the object. The image produced by this approach will be projected at the same location onto the detector due to correlated motion of X-ray source and detector which moved synchronously in opposite directions [13] [14]. This technique was used for detecting the defects size with restriction of irradiation from all directions. As required projection angle up to 200°, this technique allows to examine application of flat and large components, lack of access, limited angle accessibility, time consuming, and thick insulation or concrete coatings that normally cannot be examined with conventional projection radiography [14][15].

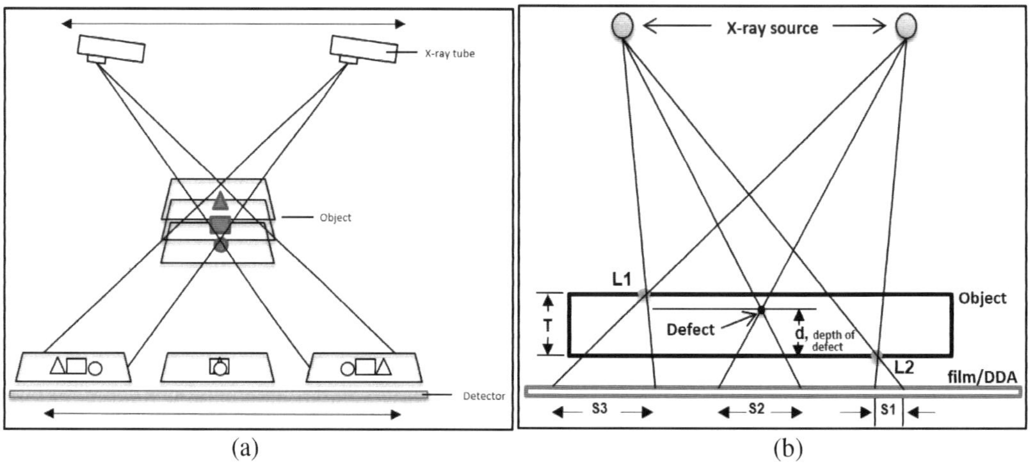

Figure 1. Basic principle of laminography (a) and lead marker method (b).

In present study, triangulation technique is performed as a simplest projection techniques in determining the depth of defects. This technique is practically required two image radiographs at two different positions of X-ray source or sample at projection angle 10° to 45°. 'Lead Marker Method' is the most method used in this technique. The schematic diagram of this method is illustrated in figure 1 (b). Measurement for depth of defects (d), can be expressed as follow:

$$d = T\frac{S2 - S1}{S3 - S1} \tag{3}$$

Where S1, S2, and S3 are the distance of shift image and lead markers while T is the thickness of the object.

In this study, two essential parameters, i.e: SR_b^{image} and SNR_N for digital laminography are studied for achievable contrast sensitivity and visibilities of defects in welded samples. Hence, regarding to these essential parameters on digital laminography technique, the preliminary experimental study and analysis were done on welded sample by triangulation technique in obtaining the high image quality the depth of defects. This study focused on cone beam configuration of X-ray system since this configuration is widely used in NDT industry [16].

2. Experimental procedure

A known defect weld test sample (SAMPLE 1) of carbon steel plate is used in this research. Welded plate with parent metal (PM) thickness 10.07mm, weld metal (WM) thickness 2.33mm, and 144 mm x 74mm in dimension. The chemical composition of the welded sample is shown in table 1. All the experimental works are conducted at Malaysian Nuclear Agency, Bangi.

Table 1. Chemical composition of low carbon steel plate.

C	Si	Mn	P	S	Cr	Ni	Fe
0.160	0.216	0.698	0.024	0.040	0.110	0.082	Balance

Study on exposure parameters of digital laminography are performed by using an YXLON X-ray tube EUO 225D model with focal spot size of 3.0mm according to EN12543. A Digital Detector Array (DDA) PerkinElmer XRD 1611 is used to acquire the image. This device has a 100 μm pixel size and sensitivity of amorphous silicon photodiodes which resulting basic spatial resolution (SR_b) of 100μm without binning. ADC resolution of DDA is 16 bits per pixel with total pixel number of 4096x4096. The Isee! Professional software from Vision in X is used for radiography image acquisition and DDA calibration. The experimental setup for digital radiography is shown in figure 2.

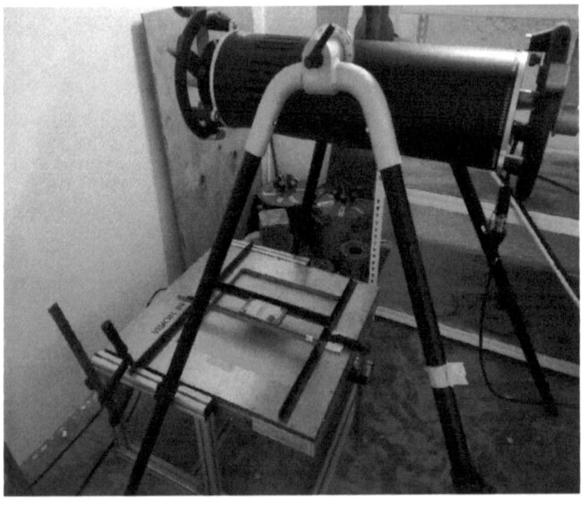

Figure 2. Experimental setup: (a) X-ray source, (b) welded sample, c) DDA.

A total of 25 experimental works are performed to study the exposure parameters by varying the current at specific voltage and acquired number of frame. The exposure details of experiment are presented in table 2.

Table 2. Exposure details.

X-ray voltage (kV)	X-ray current (mA)	Focal spot (mm)	SDD (mm)	ODD (mm)	f_{min} (mm)	No. of frame	Frame time (sec)
180 kV	0.5, 1.0, 2.0, 4.0, 6.0	3.0	700	40	263.2	1	15, 30, 60, 90

Then, geometric magnification is calculated using formula below:

$$M = \frac{SDD}{SOD} = \frac{700}{660} = 1.17 \quad (4)$$

The analysis on the Grey Values (GV) and normalized Signal-to-Noise Ratio (SNR_N) at weld metal area are measured at Region of Interest (ROI) of 20 by 50 pixels [12]. The size of the ROI is used to measure the mean intensity for reliable SNR measurements [17]. The minimum requirement for IQI and SNR_N values at weld metal (WM) are presented in table 3 [12].

Table 3. Required minimum IQI and SNR_N

Single Wire	Duplex Wire	SNR_N (WM)
W12	D8	70

IOP Conf. Series: Materials Science and Engineering **554** (2019) 012007 doi:10.1088/1757-899X/554/1/012007

A triangulation technique is performed in the preliminary study on laminography, whereas two lead markers are placed, one on the source side (L1) of the sample and the other on the DDA side (L2). These steps must be done carefully in order to avoid the images of the defects and markers do not coincide or get mixed. The detail diagram as shown above in Figure 3. The acquired radiography images are analyzed by Isee! Professional and ImageJ software in determining the distance of S1, S2, and S3. Then, the value of depth of defect, d is measured. As confirmation of this technique, another a known defects carbon steel sample (SAMPLE 2) with PM thickness of 8.00mm and WM thickness of 1.15mm is tested by using this approach.

3. Result and discussion

3.1. Evaluation of acquired radiography images

Figure 3 shows the raw and filtered acquired radiography images are captured by RT-D. The SAMPLE 1 consists of sub-surface crack approximately 30mm in length. The values of SNR_N are determined by SR_b where directly measured from the $SNR_{measured}$ as stated in (1).

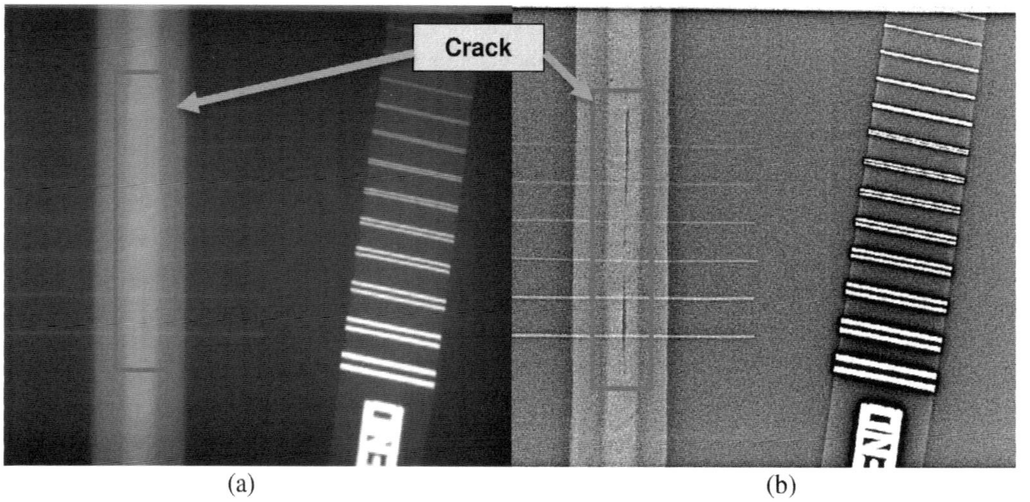

(a) (b)

Figure 3. The acquired radiography image; raw image (a) and filtered image (b).

The analysis on GV and SNR_N on all acquired radiography images are carried out with achievable contrast sensitivity and minimum image quality value. The result of single and duplex IQI wires are illustrated in figure 4.

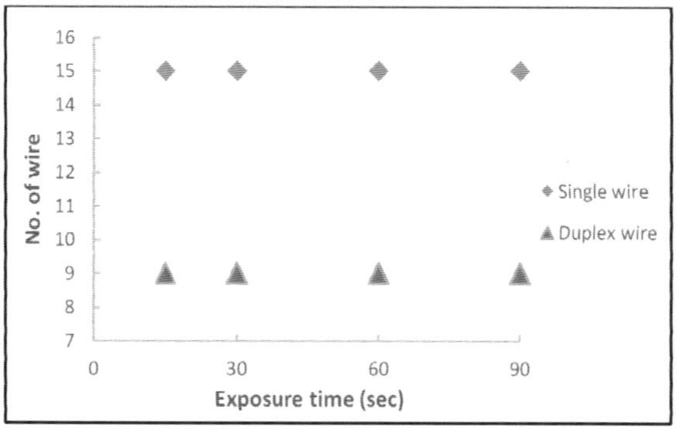

Figure 4. Minimum detectable single and duplex IQI wires.

The measurement of GV and SNR_N are plotted with the respective square root of exposure at specific frame time as shown in figure 5 (a), (b), (c), and (d).

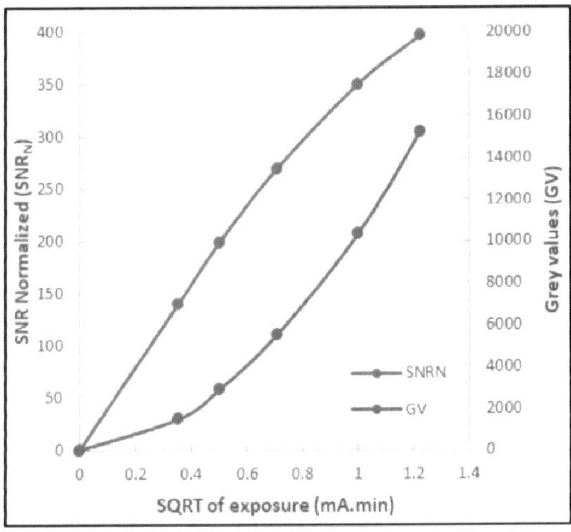

(a)

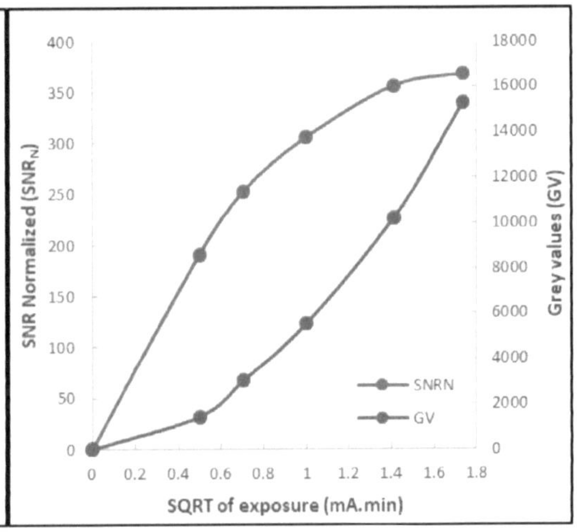

(b)

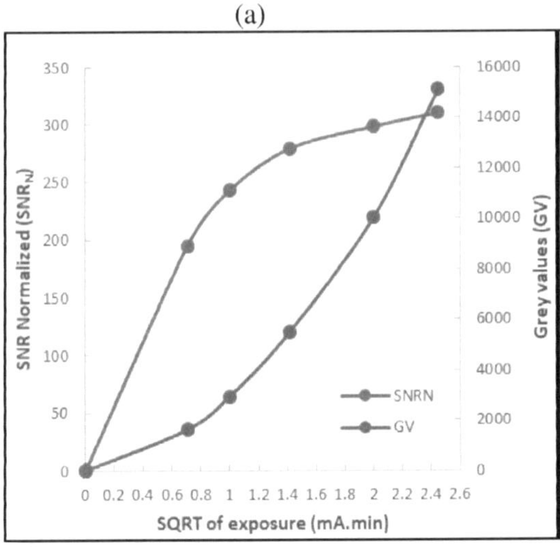

(c)

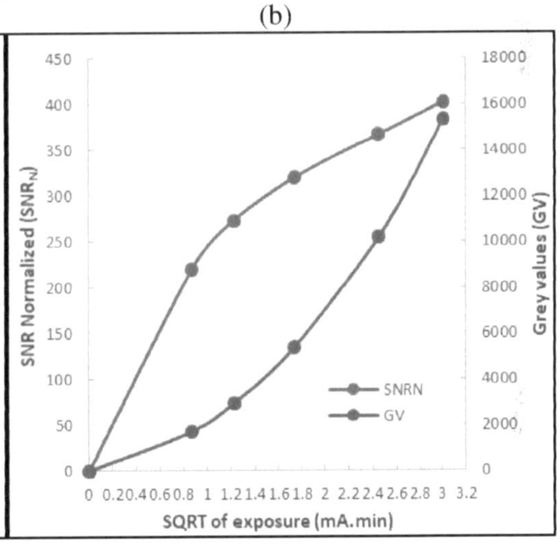

(d)

Figure 5. SNR_N and GV over the sqrt exposure of (a) frame time = 15 seconds; (b) frame time = 30 seconds; (c) frame time = 60 seconds; (d) frame time = 90 seconds.

Based on figure 5 (a), (b), (c), and (d), the GV and SNR_N increase with increasing on the square root of exposure. Increasing the radiation dose will increase radiation intensity as larger number of photons are absorbed [10]. Hence, it will produced the greater number of the SNR values with the less noise of the radiography image [18].

3.2. Depth of defect location in digital industrial radiography

According to equation 3, the relationship of the image shift and the markers are proportional to S1, S2, and S3 distance based on triangle approach. By analyzing on acquired images at two different positions, the distance of S1, S2, and S3 are measured as well as determine the depth of defect values. The determination of depth of defects (d) are mostly depends on the quality of acquired radiography image. Table 4 displays the result of the experimental work. For comparison, figure 6 shows the corresponding optical macrograph obtained by metallographic method of the SAMPLE 1.

Table 4. Result of SAMPLE 1.

PM Thickness, T	S1	S2	S3	$d_{calculated}$	d_{macro}	Error	% Error
10.07 mm	262.53mm	266.33mm	266.69mm	9.20mm	9.708mm	0.51mm	5.23%

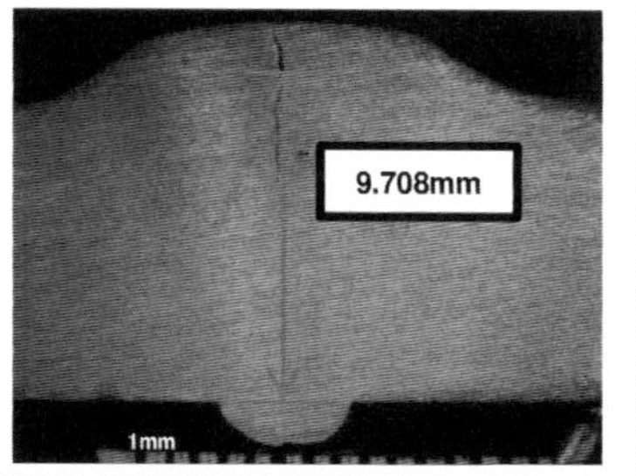

Figure 6. Macrograph image of SAMPLE 1 with defect (crack).

Further experimental work was done on SAMPLE 2 that consist of porosity and slag inclusion. The filtered acquired radiography image as shown in figure 7. The results show a good, consistent ability to determine the depth of defects to within ±1.00 mm of the actual depth. Table 5 shows the results of SAMPLE 2.

Table 5. Result of SAMPLE 2.

Defects	PM Thickness	S1	S2	S3	$d_{calculated}$	d_{actual}	Error	% Error
Porosity	8.0mm	262.99mm	264.47mm	265.62mm	4.50mm	3.95mm	0.55mm	13.92%
Slag inclusion	8.0mm	262.99mm	264.30mm	265.62mm	4.00mm	3.95mm	0.05mm	1.27%

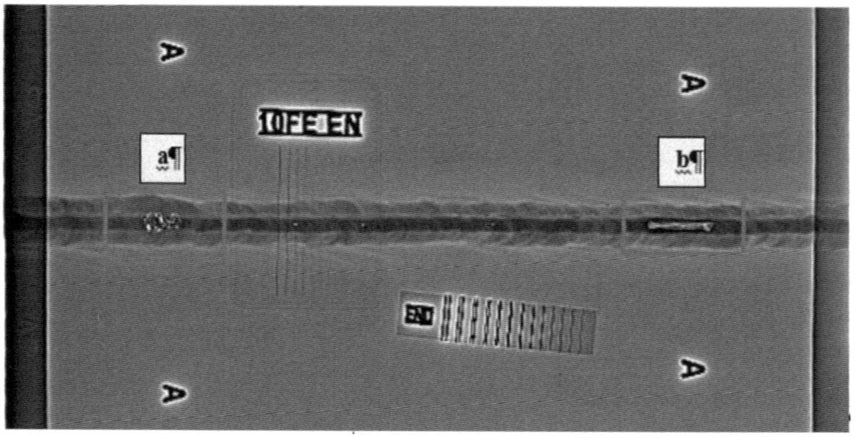

Figure 7. SAMPLE 2 with porosity (a) and slag inclusion (b).

4. Conclusion

The aim of this preliminary work is study the exposure parameters of laminography technique on carbon steel plate using RT-D. The result of optimum exposure parameters indicate that this technique is able to determine the depth of defects in carbon steel plate. Further research in digital laminography will be conducted on number of projection angle and image reconstruction algorithm technique in producing 3D image. The depth information will be discovered by applying several projection angles up to 200° projections followed by application of various image reconstruction algorithm for developments 3D image with minimum artefact inherent.

Acknowledgement
The author would like to acknowledge Universiti Kuala Lumpur, Malaysian France Institute (UniKL-MFI) and Malaysian Nuclear Agency for providing the facilities assistance in conducting this study.

References
[1] Annette O'Brien 2004 *Welding Handbook* (Miami: American Welding Society)
[2] The Welding Institute 2016 *CSWIP 3.1 Welding Inspector WIS5* (Cambridge:TWI Ltd)
[3] B M Abdullah, A Mason and A Al-Shamma'A 2013 Defect detection of the weld bead based on electromagnetic sensing *J. Phys. Conf. Ser.* **450** p 012039
[4] G Wang and T W Liao 2002 Automatic identification of different types of welding defects in radiographic images *J. NDT E Int.* **35** p 519–528
[5] U Ewert, U Zscherpel and K Bavendiek 2007 Replacement of film radiography by digital techniques and enhancement of image quality *J. NDT.net* p 4516
[6] L Lança and A Silva 2013 *Digital Radiography Detectors: A Technical Overview* (New York :Springer Science Business Media)
[7] E V Moreira, J M Barbosa Rabello, M D S Pereira, R T Lopes and U. Zscherpel 2010 Digital radiography using digital detector arrays fulfills critical applications for offshore pipelines *EURASIP J. Adv. Signal Processing* **2010 Issue 1** p 894643
[8] U Ewert, U Zscherpel and K Bavendiek 2007 *Strategies for Film Replacement in Radiography - a comparative study* IV Conferencia Panamericana de END Buenos Aires p 142
[9] BSI Standards 2013 *Non-destructive testing - Image quality of radiographs Part 1: Determination of the image quality value using wire-type image quality* (UK: BSI Standards Publication)
[10] F T Boateng, U Ewert, T Kannengiesser, U Zscherpel, A Griesche, A Kromm, S Hohendorf and B. Redmer 2016 Real-time radiography for observation of crack growth during GTA welding *Weld. World* **60** pp 931–937
[11] U Ewert and U Zscherpel 2014 Minimum Requirements for Digital Radiography Equipment and Measurement Procedures by Different Industries and Standard Organizations *11th Euro. Conf. on Non-Destructive Testing* p 460
[12] ISO 17636 Part 2 2013 *X- and gamma-ray techniques with digital detectors (Switzerland: ISO)*
[13] S Gondrom, J Zhou, M Maisl, H Reiter, M Kro and W Arnold 1999 X-ray computed laminography: an approach of computed tomography for applications with limited access *J. Nuclear Eng. and Design* **190**, pp. 141–147
[14] K Wakimoto, J Blunt, C Carlos, P J M Monteiro, C P Ostertag and R. Albert 2008 Digital laminography assessment of the damage in concrete exposed to freezing temperatures *J. Cement and Concrete Research* **38** pp. 1232–1245
[15] M Hoshino, K Uesugi, A Takeuchi, Y Suzuki and N. Yagi 2010 Development of an X-ray micro-laminography system at sPring-8 *AIP Conference Proceedings* **1365** pp. 250-253
[16] N S O'Brien, R P Boardman, I Sinclair and T. Blumensath 2016 Recent advances in X-ray cone-beam computed laminography *J. Xray. Sci. Tech.* **24** pp. 691–707
[17] BSI Standards 2005 Non-destructive testing - Industrial computed radiography with storage phosphor imaging plates (UK: BSI Standards Publication)
[18] U Ewert and U Zscherpel 2009 Industrial radiography manual of the syllabi contended in IAEA - training guidelines for non-destructive testing techniques," *Int. At. Energy Agency* pp. 2–6

IOP Conf. Series: Materials Science and Engineering **554** (2019) 012008 doi:10.1088/1757-899X/554/1/012008

Radiographic simulation and validation studies on weld joints of annular tanks and cylindrical tanks

Kumar, S.*, Menaka, M., and Venkatraman, B.

Health, Safety and Environmental Group, Indira Gandhi Centre for Atomic Research (IGCAR), Kalpakkam 603 102, Tamil Nadu, India.

Corresponding author, email: kumar257selvam@gmail.com

Abstract. The application of the radiographic simulation studies on the weld joints in nuclear fuel reprocessing tanks used for optimization of radiographic testing parameters. In the nuclear reprocessing plant, Annular and cylindrical tanks are used to store the reprocessed solution of the spent fuel pins from the reactor. These annular and cylindrical tanks were installed 15 years ago. AERB (Atomic Energy Regulatory Board) has requested for requalification of these tanks before commissioning. As part of the requalification of tanks, the volumetric NDE techniques were proposed to be used for evaluation of the weld joints in the tanks. Before carrying out the radiographic examination on the actual tanks, computer simulation of radiography was studied on standard weld pad. Radiographic simulation is utilized for various purposes in NDT. This radiographic diagnostic model utilized for study about incorporates a description of the radiation source, the interaction of radiation with test pieces and defects, and the detection process. The modeling results were validated with radiographic inspection carried out with weld pad sample of 6mm thickness. The error percentage in the optical density achieved between simulation and experiment radiographic inspection was ± 5%. Hence the optimization parameters using simulation for the weld pad helps to reduce the number of exposures, inspection time and increase defect detectability of the inspection.

1. Introduction

CORAL (COmpact Reprocessing facility for Advanced fuels in Lead shielded cell) is the pilot plant successfully demonstrated by reprocessing spent fuels from Fast Breeder Test Reactor (FBTR). Based on the feedback obtained from CORAL operations, a demonstration fuel reprocessing plant of still higher capacity has been designed and constructed. The main objectives of this plant are to reprocess the fuel of FBTR on regular basis and that of PFBR (Prototype Fast Breeder Reactor) on experimental basis [1-3]. In this plant, cylindrical and annular tanks are used to store the process solution of the spent fuel pins from the reactor. Tanks were erected 15 years ago. AERB (Atomic Energy Regulatory Board) has requested for the requalification of these tanks before commissioning. Based on AERB requirements, requalification methodology for process vessels and associated pipings erected at various cells of Demonstration Fast Reactor Fuel Reprocessing Plant (DFRP) has been planned. A sampling percentage of around 10 % covering entire spectrum of process vessels will be covered for requalification. Selection of critical process vessels to be covered in the sampling plan was decided by the designer based upon criticality, process conditions and accessibility. As part of the requalification of tanks, the volumetric NDE techniques [4] were proposed to be used for evaluation of weld joints in the tanks. Before carrying out radiographic examination on the actual tanks, computer simulation of radiography [5-7] was studied on standard weld samples in *aRTist* (Analytical Radiography Testing

Inspection Simulation Tool) software. The modelling results were validated with experimental results of weld pad samples. After that radiographic simulation was carried on actual tanks. Then optimizing the parameters was used in radiographic examination on actual tanks [8]. aRTist simulation programming is created by BAM (Federal Institute of Materials Research and Testing), Germany [9]. It consolidates diagnostic modeling of the RT inspection process with CAD oriented object description. This product tool for numerical simulation is displayed which creates practical radiographs on the premise of virtual part portrayals and very much characterized radiological parameters. This simulation tool satisfies the practical industrial requirements and connected for inspection planning. This simulation program utilizes an analytical procedure model, productive ray tracing algorithms, industry standard boundary portrayals, and a graphical UI with 3D perception to meet these requirements. This simulation procedure can be subdivided into three autonomous parts: the generation of X- ray beam, the interaction between the radiation and material, and the imaging process. The radiation material communication can be portrayed by the Boltzmann transport equation. Ray tracing model based on the attenuation law described the detailed solution of the image.For the interaction of radiation with material we limit our model to photon interactions and do not represent electrons. Appropriately the accompanying interaction mechanisms are viewed as: the photoelectric effect, coherent and incoherent scattering, and for photon energies bigger than 1 MeV pair production.

The photon transport model is using the stationary Boltzmann equation

$$\mathbf{\Omega}\cdot\nabla I(\boldsymbol{r},E,\mathbf{\Omega})+\mu(E)I(\boldsymbol{r},E,\mathbf{\Omega})=$$
$$\int_0^{\infty}\mathrm{d}E'\int_{4\pi}\mathrm{d}\Omega'\,\sigma(E'\to E,\mathbf{\Omega}'\to\mathbf{\Omega})I(\boldsymbol{r},E',\mathbf{\Omega}')+S(\boldsymbol{r},E,\mathbf{\Omega})$$

It describes the variation of the photon flux $I(\boldsymbol{r}, E, \Omega)$ at position $\boldsymbol{r}$ with direction Ω and energy E. The reduction of the flux by the interaction of photons with matter given by the linear attenuation coefficient μ (E) is the left hand side of the equation.

$$\mu(E)=\tau+\sigma_{\text{coherent}}+\sigma_{\text{incoherent}}+\pi$$

τ is the absorption coefficient, $\sigma_{\text{coherent}}$and$\sigma_{\text{incoherent}}$ are scattering coefficients and π is the pair production coefficient.The increase of the photon fluxI($\boldsymbol{r}$, E, Ω) by scattering contributions from other energiesE' and other directionsΩ' givenby the scattering cross section$\sigma(E'\to E,\Omega'\to\Omega)$ as well as by internal photon sources$S(\boldsymbol{r}, E,\Omega)$ such as X-ray fluorescence or electron-positron annihilation described in right hand side. The right hand side of the equation is vanished by neglecting the scattering contributions. Then ordinary differential equation can be directly incorporated prompting the well known attenuation law

$$I_p(\boldsymbol{r},\mathbf{\Omega},E)=I_0(\boldsymbol{r},\mathbf{\Omega},E)e^{-\int_0^{|r-r_0|}\mathrm{d}R\,\mu(\boldsymbol{r}-R\mathbf{\Omega},E)}$$

The primary photon flux I_p ($\boldsymbol{r}$, E, Ω) derived from the initial fluxI_0($\boldsymbol{r}$, E, Ω), which conveys the significant data for radiographic methods [11].

1.1. Part geometric model

The input for geometrical model is a triangulated surface model of the test piece in the industry standard STL format. Along these lines it is conceivable to work with arbitrary geometry and to interface straightforwardly with CAD programming. Material types are constantly allocated to the part geometries form an extensible material rundown within the program [9]. Initial modeling studies were carried out using stainless steel 304L weld pad of 6mm thick with dimension 100mmX100mm. Modelled weld plate in the STL format is shown in Figure 1.

Figure 1. CAD model of weld pad for simulation

1.2. Source model

With respect to X-ray beam generation in the source, aRTist proposes an X-ray tube model, which can be utilized between 30kV to 450kV.For the weld pad radiographic inspection modeling studies, 110kV, 120kV and 130 kV tube voltages were simulated. This executed X-ray tube display simulates the physical marvels engaged with bremsstrahlung and characteristic photon creation with semi-empiric model. It considers the anode angle and composition, the inherent and extra filtration and the photons leave angle. Source modeling was carried out using tungsten target with anode angle 40° and aluminum filter for reduction in beam hardening. The typical energy spectrum of tungsten target for 130KV simulated x-ray source is shown in the Figure 2. The tube current of 4.3mA was fixed for the simulation.

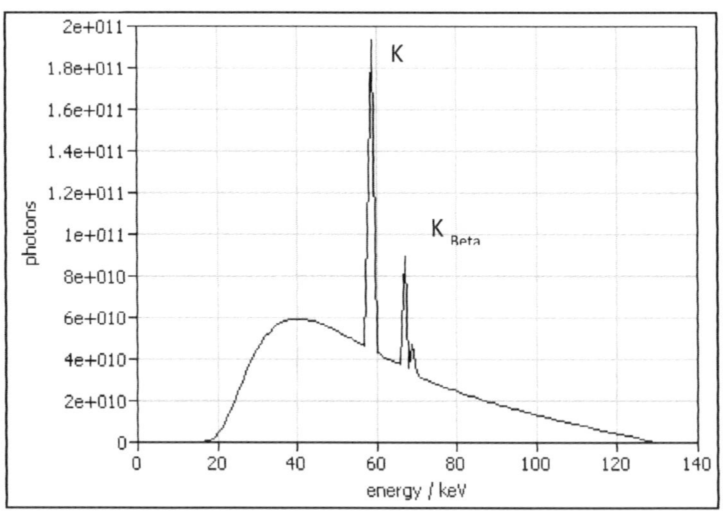

Figure 2. Energy spectra of tungsten target for 130kV

IOP Conf. Series: Materials Science and Engineering **554** (2019) 012008 doi:10.1088/1757-899X/554/1/012008

1.3. Detector modelling
Detectors are displayed in two progressive strides. The first, basic to a wide range of detectors registers the energy deposition in the delicate piece of the detector utilizing the energy absorption attenuation coefficients. The second one is particular to each kind of detector which recreates the progressive physical phenomena engaged with the energy to signal change. Different models are adjusted for the different detectors, for example, scintillating screen, saw by CCD camera, photomultiplier and a film [10]. Transmission functions depend on the property of the detectors. [11] For the modeling studies the detectors were D4 film, D5 film, and D7 film with size of 250 X105 mm.

1.4. Virtual set up
The computer simulation models the genuine review situation by characterizing a virtual a setup. Other than radiation source and detector, at least one geometrical part portrayals can be arranged freely in virtual space. Parts are modeled in CAD programming and imported into the aRTist. A cover of the parts characterized in the scene is resolved by their need. This takes into consideration adaptable defect descriptions that are autonomous from the encompassing material. To decide the attenuation of radiation the length of homogeneous material segments on the direct associations amongst source and detector indicated have been computed. Intelligent arrangeable part portrayals can without much of a stretch be overseen in the virtual 3D scene. Cover of geometries in the scene and blend by Boolean operator's prompts variable imperfection descriptions free frame encompassing host material of defect. Inside the program it is conceivable to create geometrical primitives like cuboids, ellipsoids and cylinders. Other part geometries can be given in STL format [12]. Defects such as porosity, lack of penetration, lack of fusion and cracks were embedded in the weld region with varying dimension and orientation for the study. For simulation calculation, a volumetric defect with a size of 2mm at depth 3mm in the weld region was used. In the standard 3D scene of virtual setup, the source to detector can be fixed. To optimize, the source can be moved like any other part in virtual setting. To set it at an exact distance the position may be defined numerically as in this case, it was varied from 1000mm, 800mm, 600mm. A reference point is defined using a red ball which is placed at the centre of the part to help in rotation of the part. The setup has also penetrameter models provided with the program. Figure 3 shows the simulation of radiographic testing in weld pad model with source and detector. Figure 4 shows the parameters used in the radiographic simulation.

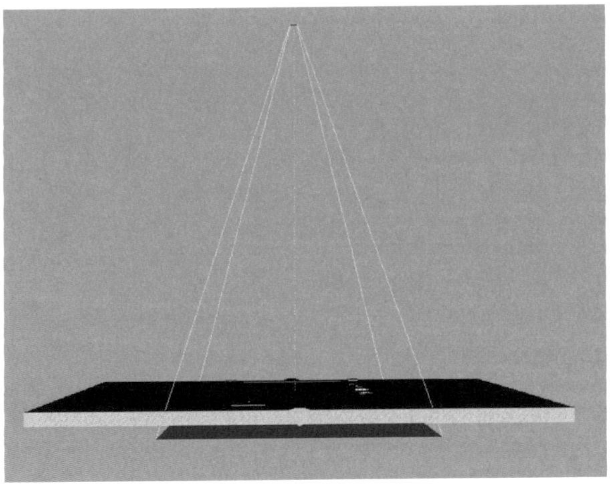

Figure 3. Simulation of Radiographic testing in weld pad model with source and detector

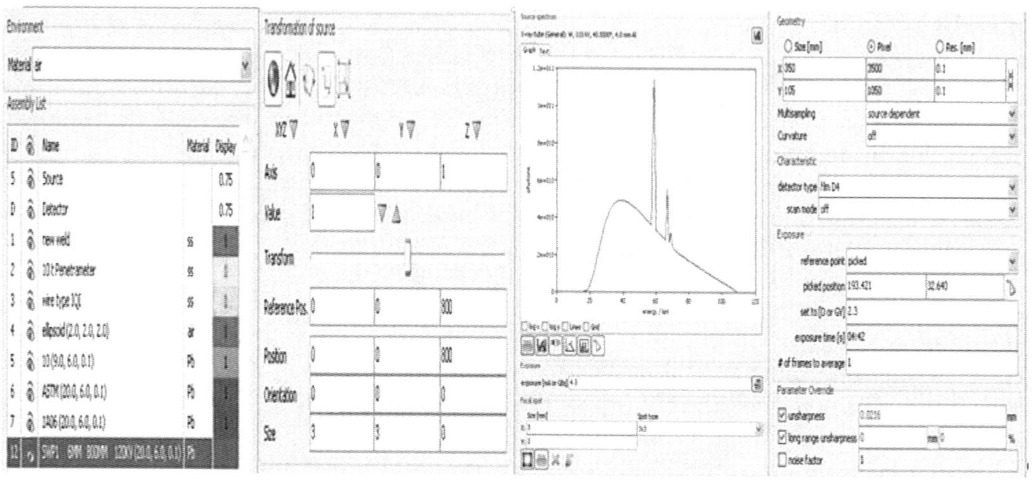

Figure 4. Radiographic testing parameters of weld pad simulation

1.5. Image computing

The last phase of image generation changes over the approaching radiation intensity to the output units of chosen detector. The detector was chosen as radiographic films from the predefined list of radiographic simulator. By default for film radiography, the exposure time is set such that the darkest point in the image has an optical density of 2.5. The darkest point with a density of 2.4 happens to be in the gas pore, while the density in the base material is around 2.3. But if we set optical density as 2.3 using a reference point in the base material, then mean density are 2.3 in the base material, approximately 2.6 in the gas pore and 2.0 in the root. Due to the film grain, the optical density is not perfectly constant but displays small variations. In the wake of setting source and detector parameters the genuine simulation of the exposure was processed. This outcome is relating to synthetic radiograph. The simulation program automatically set the correct exposure time.

1.6. Simulation flow chart for radiographic testing

Typical flow chart for the simulation is shown in below.

IOP Conf. Series: Materials Science and Engineering **554** (2019) 012008 doi:10.1088/1757-899X/554/1/012008

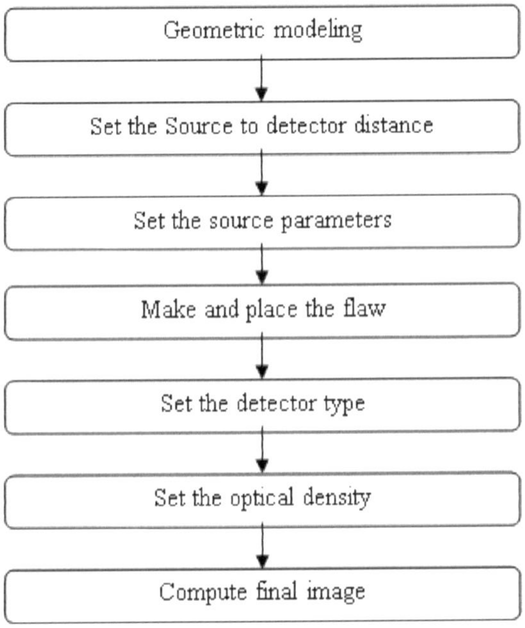

1.7. Radiographic contrast

Radiographic contrast describes the degree of differences between densities of radiograph in two adjacent areas. Contrast will be shading of gray from one extreme to the other. Radiographic contrast depends on two factors. These are subject contrast and film contrast. Figure 5 shows various parameters affecting radiographic contrast.

1.7.1. Subject contrast

It is defined as the ratio of X- ray intensities transmitted through two different areas of a specimen. It depends upon the radiation quality, intensity distribution, and scattered radiation. The higher the radiation energy, the lower would be the subject contrast.

1.7.2. Film contrast

It refers to the slope of characteristic curve of the film at a given density. It depends upon the type of film, processing conditions and optical density. It is independent of radiation quality.

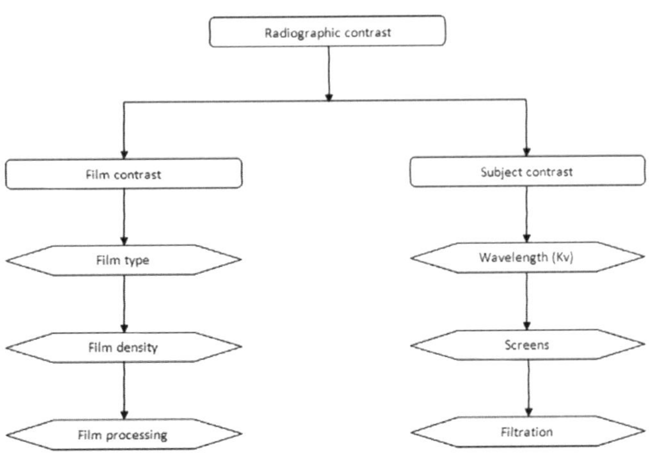

Figure 5. Radiographic contrast affecting parameters

IOP Conf. Series: Materials Science and Engineering **554** (2019) 012008 doi:10.1088/1757-899X/554/1/012008

1.8. Geometric unsharpness

Geometric unsharpness is one of the important factors used for assessing quality of radiograph. Sharpness of the image is a component of various elements. In radiographs with poor unsharpness all fine points of interest are recorded blurred. Film with a fine grain size is favored for excellent radiography, being fit for settling fine subtle elements. There is a geometrical unsharpness caused by the finite radiation source, known as the "focal spot". A large diameter source will cause a broad penumbra which implies that the edges of an image become blurred as appeared in Figure 6.

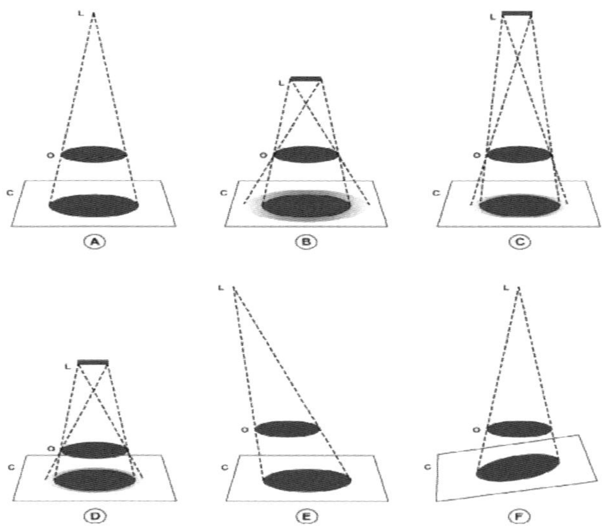

Figure 6. Geometric unsharpness affecting parameters (L- Source, O-Object, C-Detector)

The farther from the object being radiographed than the more subtle is this impact – tragically the further away the source is from the object then the more is the exposure time – double the distance quadruples the time. To limit the penumbra and increment sharpness the source should be as small as possible, the film should to be as close tothe back of the object as could be expected under the circumstances, the source should be as a long way from the object as could reasonably be expected, remembering the increasing of exposure time, and finally fine grained film should to be utilized [13]. The focal spot size of X-ray should be small, (use point source if possible). See Figure 6, A and C. The source of X-Ray should be as far away as possible from the object. See Figure 6, B and C. The detector should be as close to the object as possible. See Figure 6, B and D. The source should be directly perpendicular to the detector. See Figure 6, A and E. The detector and object planes should be parallel to each other. Compare Figure 6, A and F.

2. Experiment

Photograph of a typical weld sample and experimental setup used for validation of radiographic simulation studies is shown in the Figure 7. Stainless Steel 304L weld sample of 6mm thick with dimension of 100mm X100mm was used for the validation study. The Balteu X-ray unit used for these investigations is with current set at10 mA and maximum voltage of 160 kV. A fine grain medium speed film (Agfa D4) was used. A wire type penetrameter with wire diameters ranging from 80 microns to 250 microns was used. Plate with hole type penetrameter with step number 10 and holes 4T, 2T and 1T was also used. The penetrameters were placed on the source side is expected that both type of penetrameter would be able to indicate that the sensitivity of 2% is achieved. The radiographic parameters used for investigation of the weld pad are given in Table 1.

Table 1. Radiographic parameters employed.

Parameters	Values
Selected Voltage	110kV, 120kV,130kV
Current	4.3mA
Technique employed	Single-wall Single-image
Selected Source to Film distance	1000mm, 800mm, 600mm
Film used	Agfa D4
Penetrameter	Wire type(ASTM 1A06) and Hole type(10T)
Processing conditions	Manual, Standard
Density on radiograph	2.20 - 2.37
Sensitivity achieved	4%

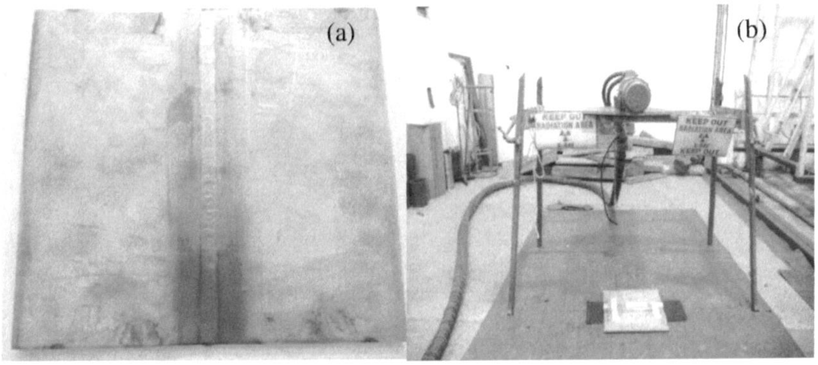

Figure 7. 6mm thick SS 304L plate TIG welded (a) and experimental setup of radiographic testing (b)

3. Results

3.1. Optical density variation

The exposure time derived from the radiographic simulation with the optical density (OD) at base material as 2.3 was used for the experimental validation. For exposure time obtained from simulation the optical density achieved in experimental study is listed for various tube voltages and source to film distance. The table also tabulates the error percentage. Optical density value for simulation, i.e. 2.3.Experiment values are shown in Table 2.

Table 2. Error percentage between experiment and simulation of optical density.

Voltage	130kV		120kV		110kV	
SFD	EXP	Error%	EXP	Error%	EXP	Error%
1000mm	2.34	+1.71	2.28	-0.87	2.21	-3.91
800mm	2.24	-2.61	2.37	+3.04	2.25	-2.17
600mm	2.35	+2.17	2.24	-2.61	2.20	-4.35

3.2. Geometrical unsharpness variation

Apart from the radiographic inspection simulation for optimization of parameters, a study was carried out on the influence of geometric unsharpness in defect sizing. Defect size valuation was carried out for the geometric unsharpness varying from 0.01 to 0.3. Table 3 shows the variation in defect size of 2mm porosity with geometric unsharpness ranging from 0.01 to 0.3. Line profile across the porosity (size 2mm) for geometric unsharpness 0.01, 0.1, 0.2, 0.3 is plotted in the Figure 8. This Figure clearly reveals that the variation in defect size is minimum for geometric unsharpness of 0.01 and it keeps increasing with increase in geometric unsharpness.

Table 3. Effect of geometric unsharpness on the defect sizing

Ug (mm)	Defect size (mm)
0.01	2.01
0.05	2.05
0.10	2.10
0.15	2.14
0.20	2.18
0.25	2.20
0.30	2.23

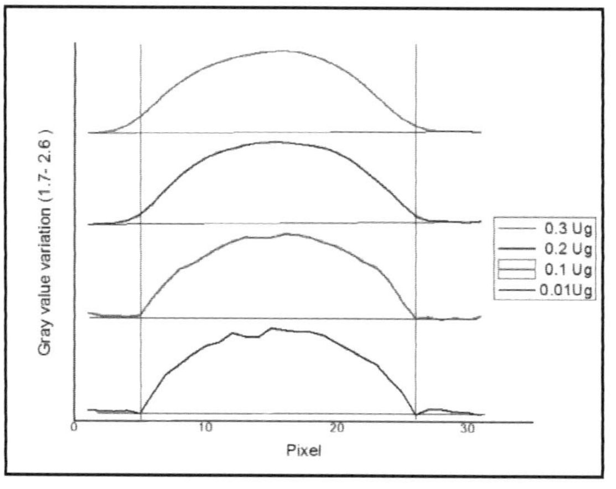

Figure 8. The line profile across the porosity with actual size of 2mm for varying geometric unsharpness

3.3. Film contrast variation

Final radiographic images were obtained by varying the source to film distance, voltage and with varying film grains for constant optical density value of 2.3. Figure 9 shows the contrast of the image obtained at different voltages. The film contrast values were computed from the radiographic images and the values are presented in Table 4.

The simulated radiographic images obtained at various tube voltages reveal the variation in the film contrast achieved and role of it in defect detection. Figure 10 shows the line profile analysis of weld in *IMAGEJ* software.

Table 4. Film contrast achieved with three possible source to film distance and tube voltage

SFD (mm)	Voltage (kV)		
	130	120	110
1000	0.42	0.52	0.62
800	0.48	0.56	0.64
600	0.50	0.58	0.69

IOP Conf. Series: Materials Science and Engineering **554** (2019) 012008 doi:10.1088/1757-899X/554/1/012008

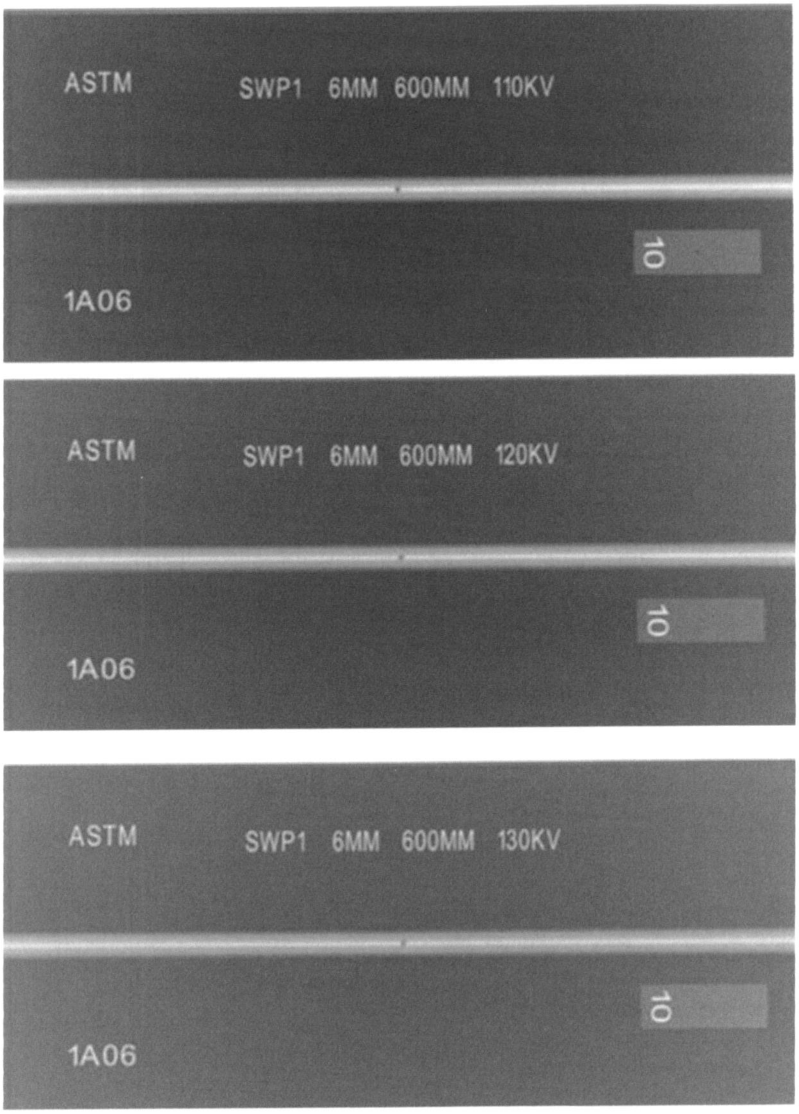

Figure 9. The synthetic radiographs of the weld pad at various tube voltage

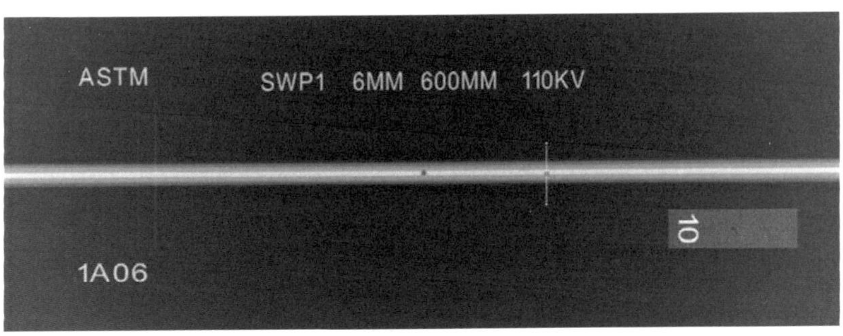

Figure 10. Analysis of the film in *IMAGEJ* software

IOP Conf. Series: Materials Science and Engineering **554** (2019) 012008 doi:10.1088/1757-899X/554/1/012008

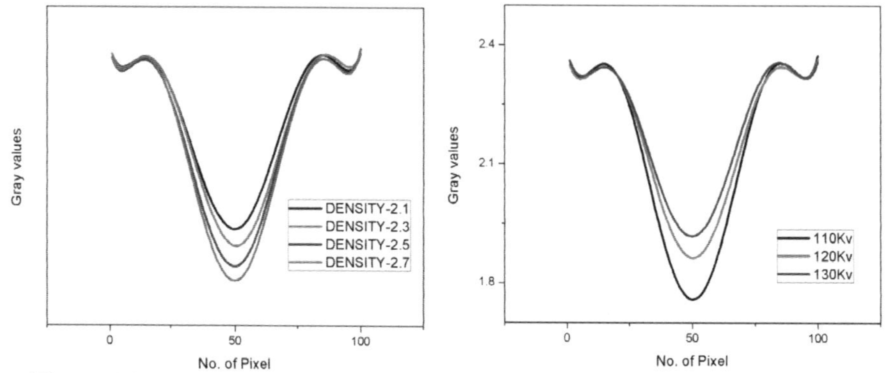

Figure 11. Contrast variation at different optical density and different voltages

Figure 11 shows contrast variation of different optical densities. SFD 600mm and D4 film are used in this arrangement for optical density value range from 2.1 to 2.7 and voltage value range from 110 kV to 130 kV. The Figure shows that high optical density give a higher contrast. Figure 11 shows gray value variation of the film along the line profile of the weld. It shows low voltage gives the high contrast. Figure 12 shows the contrast variation for different films. Here D4, D5, D7 films were used with similar SFD (600mm) and Optical density (2.3). The results clearly shows that D4 film gives the good contrast than other films.

The output of the entire process is simulated radiograph which intently coordinates experimental images. The ensuing utilization of quantitative models brings about reasonable images and coordinating exposure times.

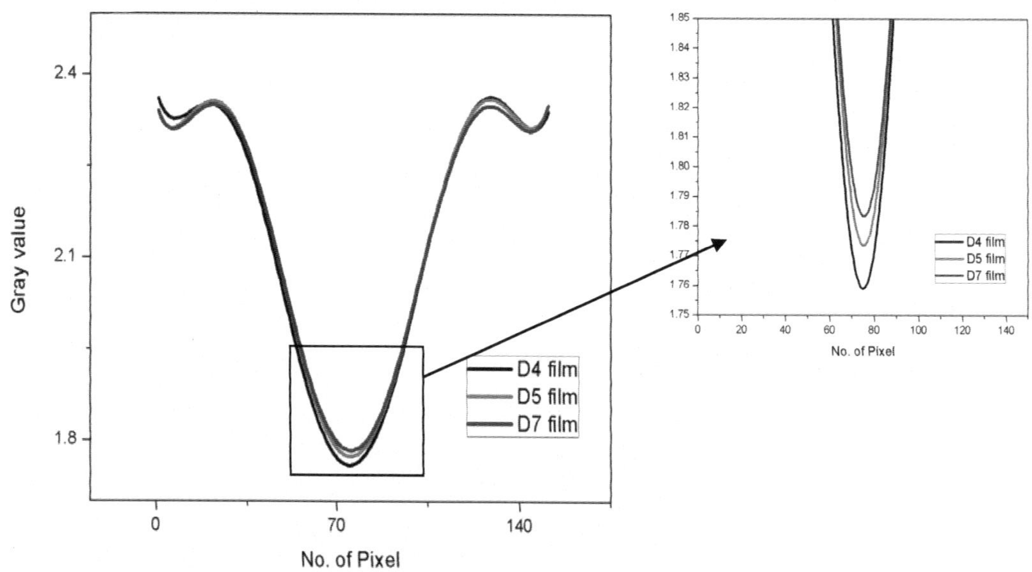

Figure 12. Contrast variation for different films (a); enlarged image showing the differences in gray values for its corresponding No. of Pixel

4. Conclusions

Simulation of radiographic examination has been carried out using *artist* software. Radiographic simulation studies were carried out to determine the optimum tube voltage, source to film distance, film density, and type of film for standard weld pad with similar configuration of the tanks. The modeling results were validated with experimental work carried out on weld pad sample of 6mm thickness. The optimum parameters obtained from the simulation for optical density of 2.3 were used for carrying out the radiographic examination of the weld pad. The optical density within range of 2.20-2.37 for different tube voltage and source to film distance were obtained in the experimental work. The error percentage in the optical density achieved by experimental radiographic inspection was ± 5% of the simulation. Hence, the optimization parameters using simulation for the weld pad helps in eliminating the over exposure & under exposure shots of the experiment, reduce inspection time and increase defect detectability of the inspection.

References

[1]. J.Kodandaraman, B.M.AnandaRao, K.Rajan, Shekhar Kumar, N.K.Pandey, A.Ravisankar, 2017, "Concurrent Trends in Indian Fast Reactor Fuel Reprocessing", International Conference on Fast Reactors and Related Fuel Cycles, Yekaterinburg, Russian Federation

[2]. Ravisankar, A., V. Vijayakumar, BM AnandaRao, U. KamachiMudali, V. Sundararaman, and R. Natarajan., 2013, "An Indian Perspective of the Development of Fast Reactor Fuel Reprocessing Technology." in International conference on Fast Reactors and related Fuel Cycles: Safe Technologies and Sustainable Scenarios FR13, Paper no. TN-CN-199/256, 4-7.

[3]. Natarajan, R., V. Vijayakumar, RV SubbaRao, and N. K. Pandey, 2015, "Experiences of Reprocessing of Plutonium-rich Mixed Carbide Fuels", *Journal of Radio analytical and Nuclear Chemistry*, **304**, 401-407.

[4]. R. Halmshaw,1987, "*Non Destructive Testing*", Edward Arnold, London, ISBN 13: 9780713136340.

[5]. Tillack, G-R., C. Bellon, and C. Nockemann., 1995, "Computer simulation of radiographic process-a study of complex component and defect geometry", review of progress in QNDE, **14A**, 665-672.

[6]. Bellon, Carsten, et al. , 1997, "Computer simulation of X-ray NDE process coupled with CAD interface." *Review of Progress in Quantitative Nondestructive Evaluation*. Springer US, 325-330.

[7]. Tillack, G-R., Christina Nockemann, and CarstenBellon. 2000, "X-ray modeling for industrial applications." NDT & E International, **33(7)**, 481-488.

[8]. Elshafiey, I., and J. N. Gray, 1996, "Optimization Tool for X-Ray Radiography NDE" ,*Journal Quantitative Nondestructive Evaluation*, **15**, 425-432.

[9]. Bellon, Carsten, and Gerd-RüdigerJaenisch, 2008, "aRTist – Analytical RT Inspection Simulation Tool for Industrial Application" In Proceedings of 17th World Conference on Nondestructive Testing, Shanghai, China.

[10]. Bellon, Carsten, Andreas Deresch, Christian Gollwitzer, and Gerd-RüdigerJaenisch., 2012, "Radiographic Simulator aRTist: Version 2." in Proceedings of 18th World Conference on Nondestructive Testing, Durban, South Africa.

[11]. Bellon, Carsten, 1997, "Computer simulation of X-ray NDE process coupled with CAD interface." Review of Progress in Quantitative Nondestructive Evaluation. Springer, Boston, MA, 325-330.

[12]. Bellon, Carsten, Andreas Deresch, and Gerd-RüdigerJaenisch., 2015, "Radiography Simulation with aRTist–Combining Analytical and Monte Carlo Methods." in Proceedings of International Symposium on Digital Industrial Radiology and Computed Tomography, Ghent, Belgium.

[13]. Raj B, Venkataraman B, 2004, "*Practical radiography*", Alpha Science Int'l Ltd, *ISBN 13: 9781842651889.*

Shearography Technique on Inspection of Advanced Aircraft Composite Material

Mohd Yusnisyam Yusof[1] , Tamil Moli Loganathan[2], Ibrahim Burhan[2], Wan Saffiey Wan Abdullah[1] , Mohd Noorul Ikhsan Mohamed[1], Ilham Mukriz Zainal Abidin, Norhayati Zakaria[2]

[1] Leading Edge Non Destructive Testing Technology, Malaysian Nuclear Agency, 43000 Kajang, Selangor.
[2] Faculty of Engineering, Universiti Putra Malaysia, 43400 UPM Serdang, Selangor.

Corresponding author, email: yusnisyam@nuclearmalaysia.gov.my

Abstract. The A320 aft fixed fairing is an advanced composite structure fabricated from laminate carbon woven and combines with honeycomb core which is externally mounted on the flap track fairing as a wing part of Airbus aircraft. These flaps are movable control surfaces to widen up the surface of the wing in order to increase the lifting. This paper discussed on the inspection of A320 aft fixed fairing by using shearography NDT technique. Shearography technique reveals the stress-affected zone due to additional loading that can be realized by the laser speckle correlation on the inspected composite panel. This A320 aft fixed fairing is identified as a scrap composite panel which having different sizes and orientations of defect. The results showed that the changes of shearing direction in laser shearography technique are an effective means to reveal the defects propagation in the composite component. The shearing at Y-axes direction is able to detect the cracks at horizontal orientation while shearing at X-axis has successfully detects the cracks at vertical orientation. Therefore, shearography could be utilized as an alternative inspection for other NDT method for inspection of composite material.

1. Introduction

Shearography technique is an optical measurement method among the new nondestructive testing NDT technique recently developed and introduced in the year of 1982 which has attracted considerable interest in a wide variety of industries including aerospace [1], oil and gas [2], automotive [3] and construction [2]. This laser based technique measures the derivative phase change difference that can be analysed with the introduction of the phase-stepping technique in the measurement system. Apart from being a non-contact and a relatively fast technique for wide area inspection [2], digital shearography have been shown to be considered reliable for detecting defects in metallic, non-metallic and also in composite materials[3,4] for qualitative inspection.

Based on Figure 1, the advanced composite of A320 aft fixed fairing component is located at wing at bottom portion, with the assembly of three main components where the upper one is known as top cover, lower is the A320 aft fixed fairing which attached in moveable fairing. Inspection of A320 aft fixed fairing component structure is increasingly important in the aerospace to make sure it is free from any manufacturing and/or service defects. At present, many NDT techniques [5,6] have been successfully employed for the detection and further analysis of the condition of advanced composite material. However, new techniques are still being explored to further improve detection and assessment capabilities in order to maintain and/or improving the quality control. Shearography could be considered as an attractive alternative for A320 aft fixed fairing inspection, since it can provide real-time inspection and visualization in a completely non-invasive way. The technique provides a highly sensitive integral measuring technique to record deformation and strain concentrations on the surface of the inspected samples or components. Therefore, with it has potential capability the application of digital shearography technique should be further explored and expanded to cater the growing needs of the aerospace industry.

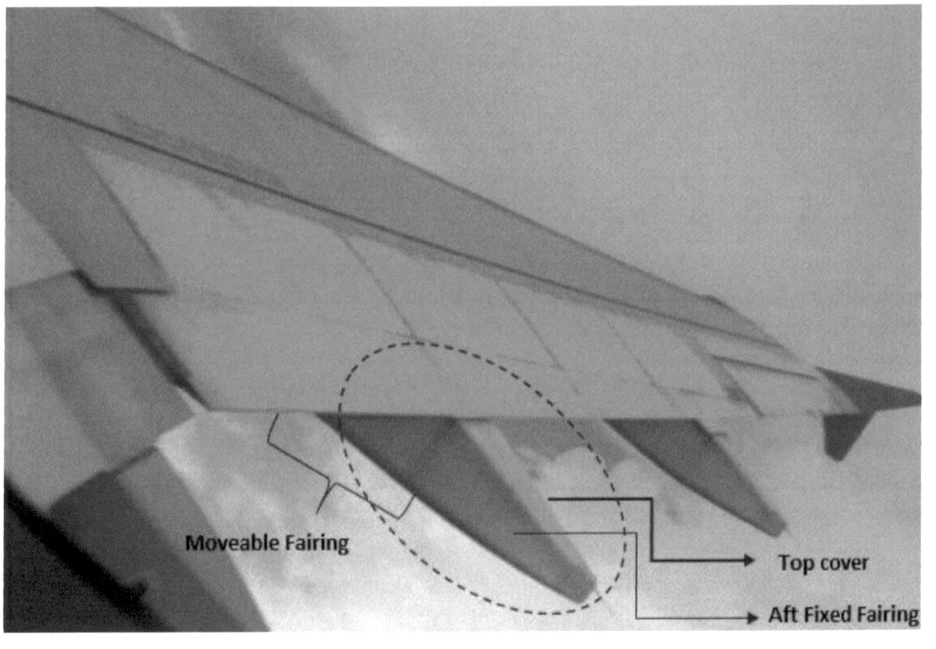

Figure 1. Wing portion of Airbus

2. Shearography: Principles of mapping, analysis and detection

The system setup of the digital shearography system used in this project work was illustrated in Figure2. It consists of a charge coupled device (CCD) video sensor as the recording media for data acquisition and digital image processing, and a Michelson shearing interferometer which employs a piezoelectric transducer (PZT) for optical phase analysis. An image processing package software, WiT7.2, is utilized with the system which assigns the intended voltage to the PZT for a four phase-stepping technique; and at the same time automatically captures and digitizes the fringe pattern image at each step. This setup is based on direct and automated measurement of phase change for phase maps representation of the displacement-derivative.

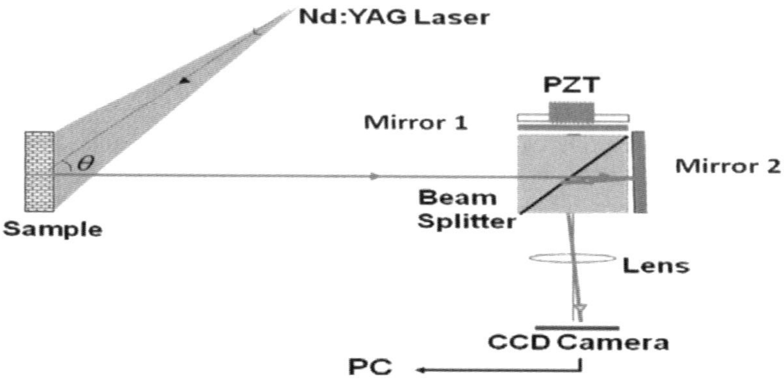

Figure 2. System setup diagram for digital shearography system

As illustrated in Figure 2, digital shearography requires the use of a laser point source for illuminating the test sample. In this project, an expanded light of Nd:YAG having a 10 mW power output and wavelength of 532 nm has been used. When the tested sample is illuminated by the expanded laser light, scattered wave fronts from an optically rough surface are sheared and interfered on the CCD sensor forming speckles as shown in Figure 3.

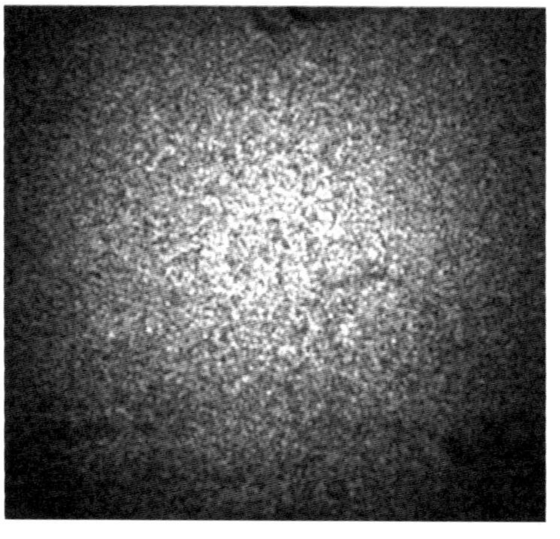

Figure 3. Laser speckle obtained by the CCD camera

The simplified mathematical equation of the intensity of the laser speckle detected on every pixel of the CCD is [5]:

$$I = A + B\cos(\phi) \tag{1}$$

Where ϕ represents the random phase difference between scattered wave front from two points on the sample surface, A is the background brightness and B is the modulation term of interference.

When the sample is applied with stress which cause it to deform slightly, both A and B could be regarded as constants. Furthermore, a relative change of surface displacement from the two points will produce a relative change of optical phase difference Δ, while the rest of Eq. (1) remains unchanged. Therefore, image intensity function after applying stress can be written:

$$I' = A + B\cos(\phi + \Delta) \tag{2}$$

With the subtraction of Eq. (1) and Eq. (2) of the two digital images and displaying the absolute value, visible fringes embedded with random speckle will be displayed. Figure 3 illustrates an example of the fringed pattern obtained by the subtraction of Eq. (1) and Eq. (2) for a pressure vessel sample deformed due to a central point load conducted in our previous work [5].

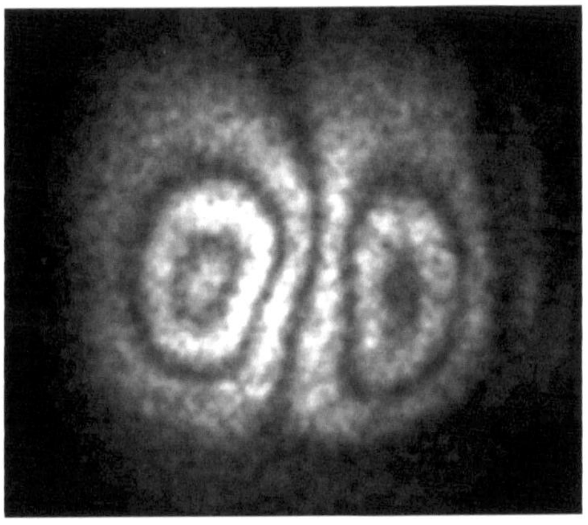

Figure 4. The fringe pattern of a pressure vessel with deformation due to a central point load

With the introduction of the phase-stepping technique, automated fringe-phase interpretation becomes feasible and the displacement derivative for the deformation can be represented by phase maps. Eq. (1) and Eq. (2) contain four unknowns A, B, ϕ and Δ, while I is the digitized pixel intensity recorded by the camera. The intention of applying the phase-stepping is to superimpose known phases using whatever means to generate a system of equations to solve for the unknown Δ. For practical reason, phase-stepping of four steps are introduced as 0, $\pi/2$, π, $3\pi/2$ between four separate frames, hence Eq. (1) becomes [6]:

$$\left.\begin{aligned} I_1 &= A + B(\cos\phi + 0) \\ I_2 &= A + B\left(\cos\phi + \frac{\pi}{2}\right) \\ I_3 &= A + B(\cos\phi + \pi) \\ I_4 &= A + B\left(\cos\phi + \frac{3\pi}{2}\right) \end{aligned}\right\} \tag{3}$$

The initial random phase ϕ is determined using the following equation:

$$\phi = arctan\frac{I_2 - I_4}{I_3 - I_1} \tag{4}$$

The same process is repeated for the deformed speckle image using Eq. (2)

$$\phi + \Delta = \arctan \frac{I_2^{'} - I_4^{'}}{I_3^{'} - I_1^{'}} \tag{5}$$

The phase change distribution due to displacement-derivatives is subsequently obtained by subtracting Eq. (4) from Eq. (5). Since the phase-step was introduced between each captured frame separated in time, the phase change Δ is wrapped in the range of $-\pi$ to π . However phase unwrapping is necessary to recover the true phase map. Both wrapped and unwrapped phase of Figure 3 is shown in Figure 4. Localized out-of-plane displacement anomalies will manifest itself as fringe condensation and can be further analyses for defect detection and characterization. The main benefit and purpose of presenting and analysing the phase images is that it can verify the obtained images from digital shearography inspection are from the stressed area and not from background noise during the test. Furthermore, quantitative evaluation is possible with further analysis of the displacement-derivatives of the unwrapped phase images; though knowledge about fringe pattern and wrapped images are essential for accurate analysis

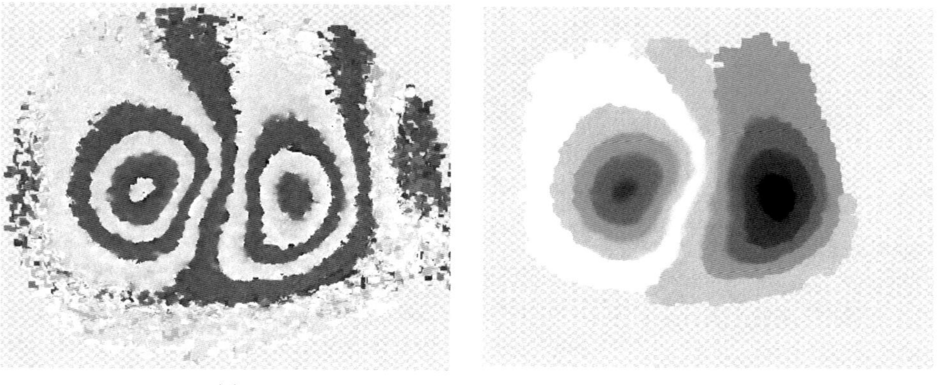

(a) (b)

Figure 5. Phase images obtained by the phase stepping technique for the pressure vessel sample shown in Figure 4, (a) wrapped phase image, (b) unwrapped phase image

3. Inspection Investigation of A320 Aft Fixed Fairing

Based on the investigation by shearography for inspection of A320 aft fixed fairing in an advanced composite sample, a 500 Watt spotlight was used as the stress mechanism applied at the front surface of A320 aft fixed fairing sample. The distance between the spotlight and the aft fixed fairing sample during the heating (5 and 20 second) of the sample was position at 300 mm; the distance and heating period where the whole surface of the sample can be optimally heated as shown in figure 6. In this inspection, shearography shows the effect of a local strain concentration due to stress loading if a crack present, on the visible surface of this sample. The A320 aft fixed fairing to be inspected is externally loaded by heating the A320 aft fixed fairing, this will produce the external stress of the A320 aft fixed fairing surface. In the areas of weakened surface caused by flaw, the rear surface of the sample is deformed more than the areas of without flaw. The phase change difference of a stress loaded and unloaded state provides information about the deformation of the rear surface of the A320 aft fixed fairing.

IOP Conf. Series: Materials Science and Engineering **554** (2019) 012009 doi:10.1088/1757-899X/554/1/012009

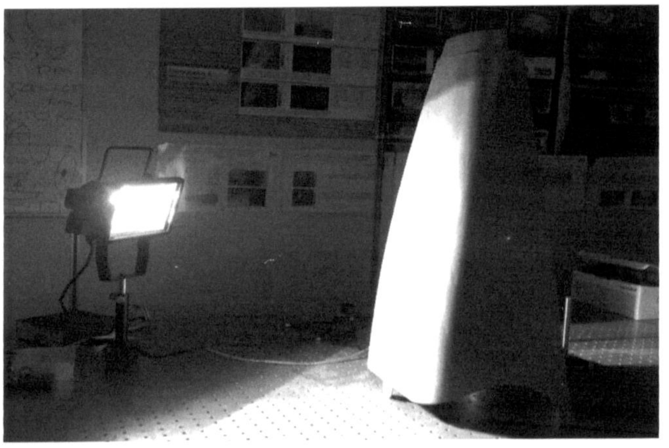

Figure 6. Arrangement of 500Watts infrared spotlight during heating of A320 Aft Fixed Fairing sample

The inspection is done at X and Y-axes shearing direction of shearography. Shearing produced by the Michelson configuration, leads to two image shearing, originated from two mirrors that are fixed orthogonally with the beam splitter. The shearing direction is changed by tilting or moving these two mirrors to the different angle. In this case two images interfere with one and the other by given 10 mm amount of shearing on the image plane as shown in figure 7. These two images should have the same intensity distribution provided that the light division by the beam splitter is an equal amount to each mirror.

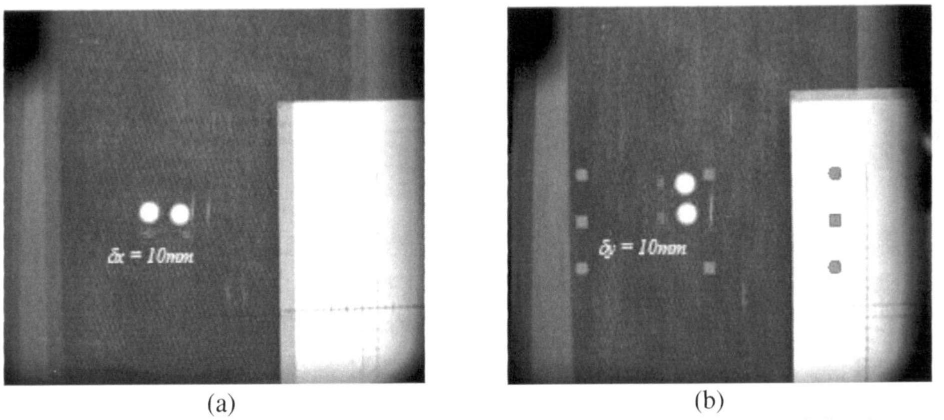

(a) (b)

Figure 7. Two images interfere with one and the other by 10 mm amount of shearing at (a) X- Axes shearing and (b) Y-Axes shearing of A320 Aft Fixed Fairing sample

The A320 aft fixed fairing sample has the thickness of 30 mm lamination of carbon woven with honeycomb structure inside the sample, as shown in Figure 8. As it can be seen in Figure 8(a) front surface and Figure 8(b) back surface with multi direction cracks in honeycomb structure of A320 aft fixed fairing Sample. The presence of the crack is intended in the inspection as a reference defect of A320 aft fixed fairing location. The schematic illustration of cracks size measurement is shown in figure 9. The measurement indicates that the higher size of crack is 3.0 cm while the lowest size is 0.8 cm.

IOP Conf. Series: Materials Science and Engineering **554** (2019) 012009 doi:10.1088/1757-899X/554/1/012009

Figure 8. (a) front surface of A320 Aft Fixed Fairing sample

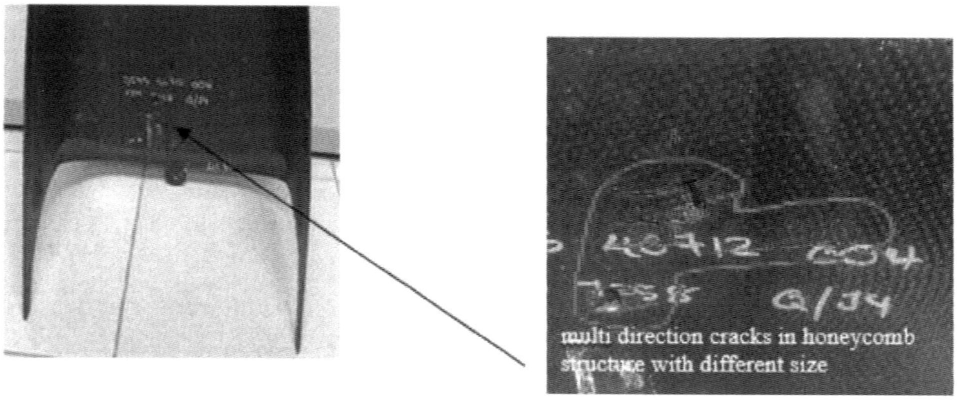

Figure 8. (b) back surface of A320 Aft Fixed Fairing sample

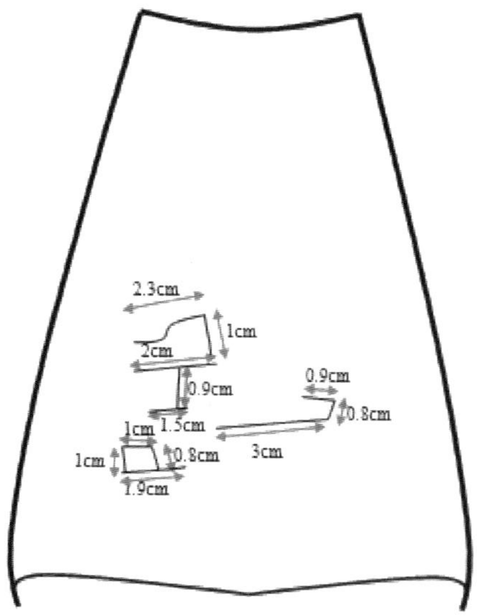

Figure 9. Schematic Illustration of cracks size in A320 Aft Fixed Fairing sample

IOP Conf. Series: Materials Science and Engineering **554** (2019) 012009 doi:10.1088/1757-899X/554/1/012009

The shearography results for the deformation of the aft fixed fairing sample surface by the applied stress by the spotlight are shown in Figure 10. The figures show the fringe patterns obtained at X and Y-axes shearing direction of shearography inspection.

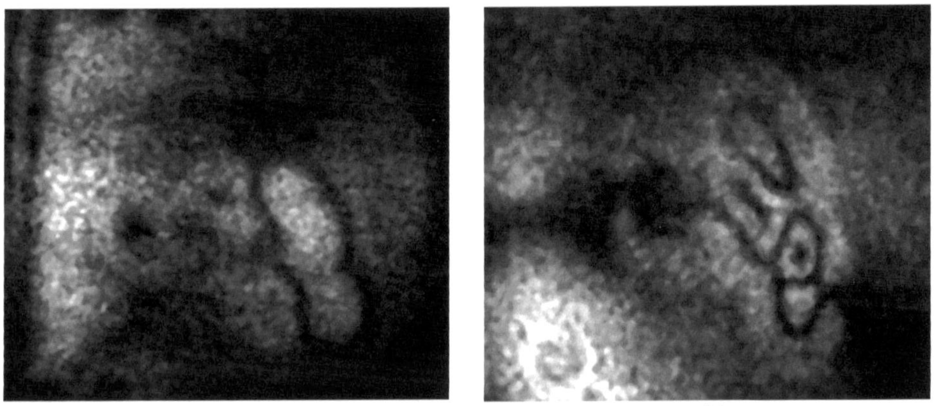

(a) (b)

Figure 10. shows fringe pattern result of A320 Aft Fixed Fairing sample (a) X- Axes shearing and (b) Y-Axes shearing

The consecutive wrapped phase representation of the displacement-derivative from the obtained fringe patterns is shown in Figure 11. The figure evidently shows a strain concentration indication due to the presence of cracks. This relatively high strain concentration is due to the higher expansion of the crack compared to the A320 aft fixed fairing the applied stress. Due to the unsmooth condition of the A320 aft fixed fairing sample surface, a smeared effect of the strain distribution can be observed with the obtained figure. However, as the strain concentration indication is obvious in the figure, the presence of the cracks can be instantaneously identified.

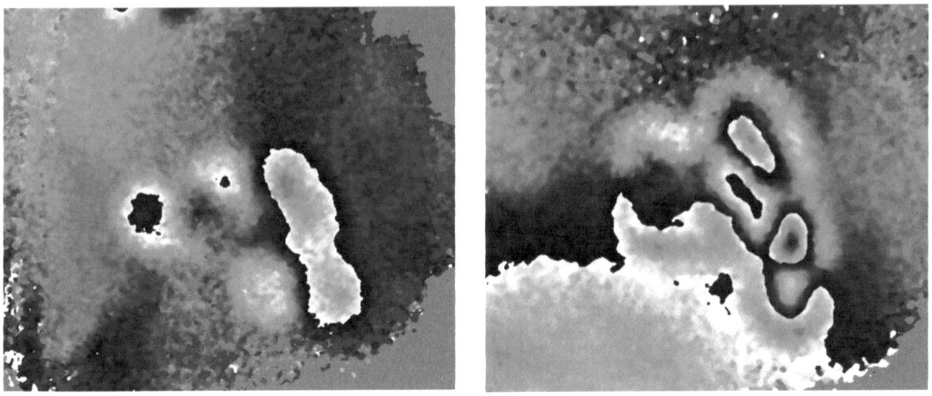

(a) (b)

Figure 11. shows wrapped phase result of A320 Aft Fixed Fairing sample (a) X- Axes shearing and (b) Y-Axes shearing

Analysis of the true phase map by the unwrapped phase image of the displacement-derivative presentation can be conducted to further verify the presence of cracks in the A320 aft fixed faring sample. Figure 11 shows the unwrapped phase representation at the A320 aft fixed faring sample surface. From the Figure 12, a clear and visible indication presence of crack can be observed.

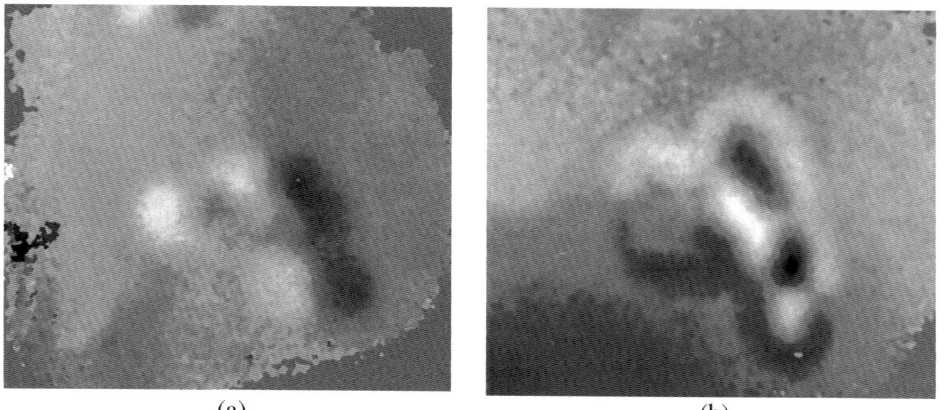

(a) (b)

Figure 12. shows unwrapped phase result of A320 Aft Fixed Fairing sample (a) X-Axes shearing and (b) Y-Axes shearing

4. Conclusion

This paper has reported the application of shearography for the inspection of A320 aft fixed fairing of advanced composite material. Inspection investigation has been carried out using a spotlight as a stress mechanism and utilizing a four phase-stepping technique in producing the qualitative results of fringe pattern and wrapped phase results for the deformation of the cracks in A320 aft fixed fairing sample. The qualitative results have shown that the cracks in the A320 aft fixed fairing sample can be detected and instantaneously identified through the strain concentration indication of the displacement-derivative by shearography inspection technique. Future work will concentrate on further evaluation of the technique and the construction of image and data processing algorithm for crack evaluation in the composite structure via the digital shearography inspection technique.

References

[1] Steinchen W, Yang L, Kupfer G and Mackel P (1998) Non-destructive testing of aerospace composite materials using digital shearography *Journal of Aerospace Engieering* 212 **1** 21-30

[2] Abdullah WSW, Yusof MY, Puad ANA and Hoque ME Analysis of corrosion in pressurized pipelines by advanced laser shearography (2009) *Engineering e-Transaction* 4 **1** 51-55

[3] Santos F, Vaz M and Monteiro J (2004) A New Set-Up for Pulsed Digital Shearography Applied to Defect Detection in Composite Structures *Optics and Lasers in Engineering* 42 **2** 131-140

[4] Kalms M and Olsten M (2003) Mobile Shearography System for the Inspection of Aircraft and Automotive *Components Optical Engineering* 42 **5** 1178-1518

[5] Hung YY, Chen YS, Ng SP, Liu L, Huang YH, Luk BL, Ip RWL, Wu CML and P.S. Chung. (2009) Review and Comparison of Shearography and active Thermography for Nondestructive Evaluation *Material Science and Engineering* **64** 73-112

[6] Miller G, Gaydecki P, Quek S, Fernandes BT and Zaid MAM (2003) Detection and Imaging of Surface Corrosion on Steel Reinforcing Bar Using Phase-Inductive Sensor Intended for use with concrete *NDT & E International* 36 **1** 19-26

IOP Conf. Series: Materials Science and Engineering **554** (2019) 012010 doi:10.1088/1757-899X/554/1/012010

Ultrasonic Evaluation at the Interface of Aluminum/Steel Obtained by TIG Welding

Qumrul Ahsan[1] & Hiroshi Kato[2]

[1]Carbon Research Technology Group, Advanced Manufacturing Centre, Faculty of Manufacturing Engineering, Universiti Teknikal Malaysia Melaka, Hang Tuah Jaya, 76100 Durian Tunggal, Melaka, Malaysia.
[2]Department of Mechanical Engineering, Division of Mechanical Science and Engineering, Saitama University, 255 Shimo-Okubo, Sakura-ku, Saitama-city, Saitama 338-8570, Japan

Corresponding author, email: qumrul@utem.edu.my

Abstract. Aluminum filler (Al-5%Si) is bonded on steel substrate by gas tungsten arc welding process under different current setting conditions. After welding a thorough metallographic analyses using optical and scanning electron microscopy were carried out to observe Al-Steel interface. Ultrasonic measurements by immersion and scanning acoustic microscopy were made to characterise the bond at Al-Steel interface. The ultrasonic wave reflected from the Al-Steel interface clearly indicated the more debonded area when the welding current is changed from direct current straight polarity (DCSP) mode to alternate current high frequency (ACHF) or the welding current in case of current for DCSP mode is reduced. This may be attributed either adequate amount of oxides along with intermediate phase make the bond interface rough and uneven due to introduction of ACHF mode or debonded areas are higher due to lower heat input when the current reduced to 125 amps from 150 amps in DCSP mode. Whereas smooth and even bond interface due to uniform heat input resulted stable Al-Steel interface and the mapping of bonded area by ultrasonic measurement showed spectrum of minimum peak for using current of 150 amps in DC mode. Thus the non-destructive assessments of interface between dissimilar materials through ultrasonic measurements could be important tools for evaluating the state of the bond.

1. Introduction

Aluminum and its alloys are widely used in fabrications because of their lightweight, good corrosion resistance and weldability. The unique combination of light weight and relatively high strength makes aluminum the second most popular metal that is welded. Light weight engineering structures are the prime concerns for automobile and shipbuilding industries. To reduce the weight of the structural components the combination of steel and aluminum alloy fabrication is considered to be an efficient measure. Several welding methods viz. explosive bonding or rolling [1] solid-phase bonding methods, such as friction welding [2], ultrasonic joining [3], rolling [4] and laser method [5, 6] are being utilized to join steel and aluminum. The investigations made by Yan eta. [5] clearly indicated the microstructural segregation at the interface of steel-Al joints clearly produced Fe-Al intermetallic compounds (IMCs), such as Fe_2Al_5, $FeAl_3$ and FeAl, that may improve the hardness of the interface layer dramatically but reduce and ductility of weldment with increased susceptibility of solidification cracking in the fusion zone. Studies on friction stir welding (FSW) [7-11] for steel/aluminum weld have also been reported and the results show that the FSW joints exhibit superior mechanical and chemical properties compared to gas tungsten arc welding (GTAW) joints due to the presence of finer grains the weld metal of FSW joints. Many attempts have also been made so far to deposit aluminum alloy on commercial grade steel substrates by fusion welding process [12, 13]. Studies from Pouranvari et.al [12] indicated that low heat input rate of GTAW process reduce the growth of intermetallic compounds during joining. In a different study Su et. al. [13] prepared Al–steel lap joints from direct-current pulsed gas metal arc welding (DPG) and alternate-current double-pulse gas metal arc welding (ADG) and showed that ADG process offers higher joint strength than that offered by DPG process due to thinner intermetallic compounds layer resulted from lower heat input.

However, sound joints are difficult to produce, because hard and brittle intermetallic compounds are formed at the weld whenever steel is welded to aluminum by fusion welding. The reason for this is attributed to the large difference between their melting points (933 K for Al and 1811 K for Fe), the very few solid solubility of iron in aluminum, and the formation of brittle intermetallic compounds such as Fe_2Al_5 and $FeAl_3$ [14]. Further, differences in their thermal properties like expansion coefficients, conductivities, and specific heats lead to internal stresses after fusion welding. In addition, presence of oxide layers during welding at steel aluminum interface can cause debonding. Therefore, fusion welds of iron and aluminum suffer from cracking with brittle failure in service [14]. Quality control of steel-aluminum weldment is considered as the prime concern for the fabrication industries. Tools and techniques especially utilization of ultrasound in non-destructive evaluation is widely used to ensure the proper quality of the weldment.

Cracks, formation of pores and other discontinuities in deposit are also of great concern to determine. A very efficient method for detecting these defects and measuring layer thickness is the ultrasonic method [15] and this can lead to strong effects on the ultrasonic properties. Hudgell [16] gave a description of austenitic clad components and of the ultrasonic testing of such components. Investigations into the energy aspect of the reflectivity for the lamination-type flaws show that the energy reflected depends on the specific wave resistance of the layers involved and the kind of substance filling the space between them, as well as on the ratio of flaw thickness and ultrasonic wave length. Generally, in the case of flaws of delamination type, the reflectivity increases and this increase is used as the basis for the ultrasonic flaw detection methods.

In the present work, ultrasonic measurements were carried out to evaluate the bond interface of steel and aluminum at different welding conditions. Welding in steel substrate with Al-5% Si alloy as filler is manually carried out at different current settings under constant welding speed through welding torch. Studies have been carried out to obtain the relationships between the state of the bonding interface and ultrasonic parameters, such as the transmissibility (or reflectivity) of an ultrasonic wave at the bonding interface and an ultrasonic waveform traveling through the interface.

2. Experimental

2.1. Preparation of specimen

In this study, steel plates were used to make weld bead deposit of Al alloy on it. The chemical analysis of the plate was carried out by optical emission spectroscopy (OES) and the result found from OES is given in Table 1. From the OES result, it was found that the steel plate is simple grade of mild steel. The mild steel plates were machined to the required dimensions of 75 mm × 50 mm × 8 mm. First the contaminants and oxide film on the steel surface were removed by steel brush. Final cleaning of the steel faying surface was carried out by acid pickling using 5% hydrochloric acid for 10-12 minutes.

Table 1. Chemical composition in wt.% of steel plate and Al filler used in this investigation. (OES)

Materials	Elements (in wt. %)					
	C	Si	Mn	Cu	Fe	Al
Mild Steel Plate	0.13	0.09	0.5	<0.01	bal	
Al-4043 filler	-	4.5 - 5.0	-	<0.3	-	bal

An A4043 (Al-5%Si) grade filler rod was used for gas tungsten arc welding (GTAW) process, and single pass welding was used to deposit aluminum weld bead on steel surface. The welding procedure was carried out by a certified welder. The plates were clamped and welded manually in the flat position using Al-5% Si filler metals at different welding currents with constant welding speed. High purity (99.99%) argon shielding gas with a flow rate of 15 liter/min was used in all welding trials. The welding current modes for ACHF and DCSP with welding parameters are presented in Table 2 to fabricate aluminum-steel (Al/Steel) clad in this study. Usually DC welding is the preferred type of welding either using direct current straight polarity (electrode negative) or direct current reverse polarity (electrode positive) to produce a smoother weld than AC. As DC delivers a constant and consistent current it is expected that welding of aluminum on steel substrate may produce a smoother interface with thin intermetallic layer. Whereas, when using AC, the current swings with changing polarity and causing fluctuation in arc which may disrupt the aluminum steel interface with higher chance of making the bond interface rough and uneven. The welding heat input rate (Q_w) for both DCSP and ACHF was calculated using following formula [17]:

$$Q_w = \frac{\eta VI}{S} \tag{1}$$

where, V is the arc voltage, I is the welding current, S is the welding speed (measured as the welding length divided by the welding time) and η is arc efficiency (considered as 0.6, a typical value for GTAW process [17]).

Table 2. Welding conditions and process parameters

Sample ID	Welding Process	Welding Mode	Current Freq. (Hz)	Average Voltage (V)	Average Current (A)	Welding Speed S (mm/min)	Heat Input Q_w (kJ/mm)
ACHF 150	GTAW	ACHF	50	15	150	60	540
DCSP 150	GTAW	DCSP	-	15	150	60	540
DCSP 125	GTAW	DCSP	-	15	125	60	450

After welding, the welded plates were cooled in air. Aluminum weld deposit for each sample was machined by milling cutter to make it around 2 mm thick. The steel substrate of each sample was also machined to make smooth and parallel surface with aluminum surface, and the thickness of the steel was made to around 7 mm thick.

2.2. Ultrasonic measurement

In order to characterize the aluminum-steel interface, the ultrasonic measurement by immersion method and the scanning acoustic microscope along were used. An experimental setup of ultrasonic measurement by immersion methods is shown in Fig. 1. The transducer of 10 mm diameter generating a longitudinal wave of 20 MHz in frequency with a focal distance of 25.4 mm in water was used. A distance between the transducer and the specimen (water path) was controlled so that the ultrasonic

IOP Conf. Series: Materials Science and Engineering **554** (2019) 012010 doi:10.1088/1757-899X/554/1/012010

wave was focused at the steel–aluminum interface. In each sample, the ultrasonic measurement was carried out at 10 x 10 scanning points in an area of 9 x 9 mm with a pitch of 1 mm by irradiating the ultrasonic wave from steel side, whereas scans were made on 20 points in a linear distance with a pitch of 1 mm from the aluminum side for same sample. In this method, the probe was kept constant distance from the surface of the object to be inspected, and scanned the surface to detect irregularities of the interface.

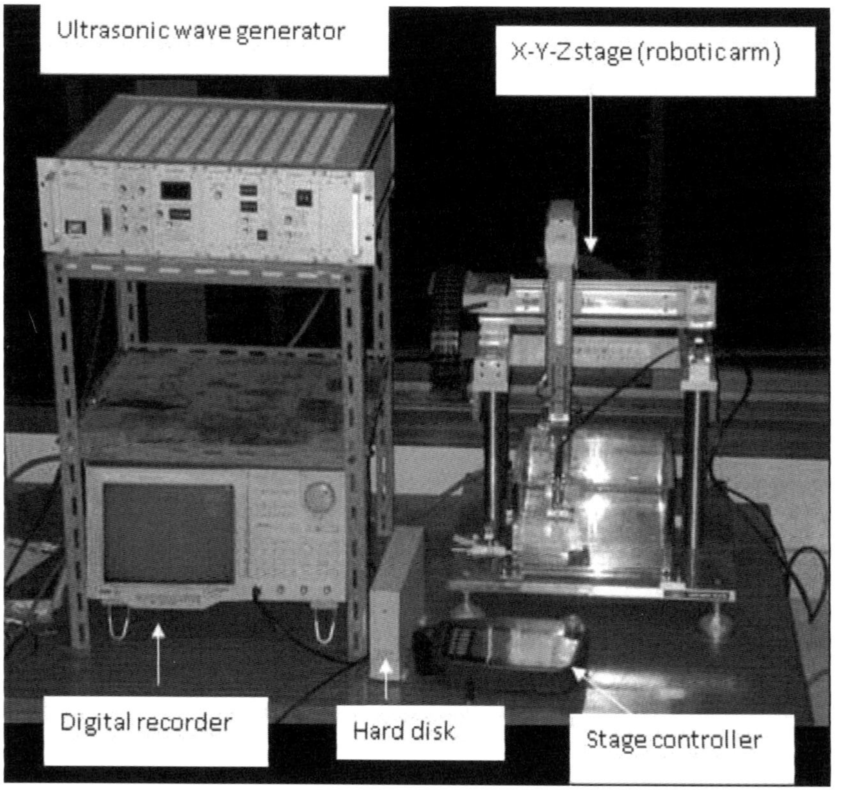

Figure 1. Experimental set up of ultrasonic C-scan system for immersion scanning.

Scanning acoustic microscope (SAM) as shown in Fig. 2 was used to obtain acoustic intensity image from the aluminum side with a probe generating a wave of 50 MHz in frequency with a focal distance of 10 mm in water to produce high resolution images at the aluminum-steel interface, and interfacial separation and irregularities at weld interface were detected. In the present analysis, the maps of an area of 18 mm x 18 mm were obtained at six different locations and combined for each sample. The absolute value of the highest amplitude signal within the gated region was recorded.

IOP Conf. Series: Materials Science and Engineering **554** (2019) 012010 doi:10.1088/1757-899X/554/1/012010

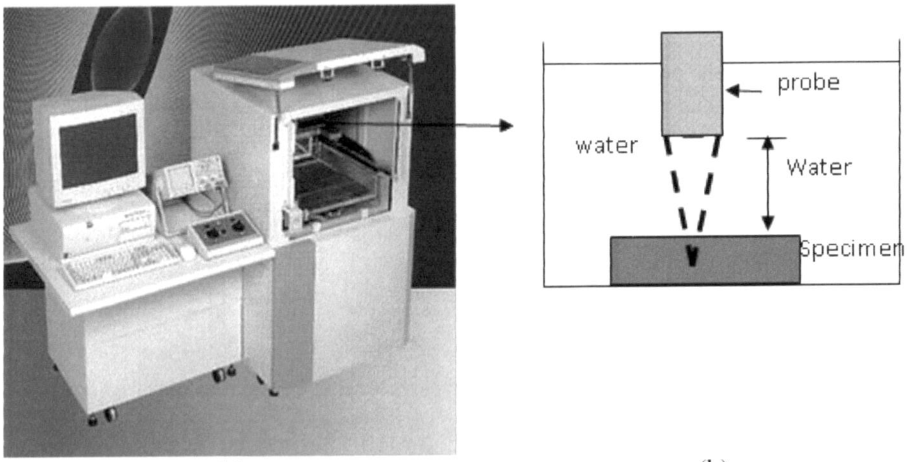

(a) (b)

Figure 2. (a) Scanning Acoustic Microscope and (b) specimen and probe position in the water.

2.3. Metallography

After ultrasonic inspection, samples were cut into small sections. Then the sections were mounted in epoxy resin under pressure. Optical metallography was performed on all specimens. Before metallography, specimens were longitudinally sectioned and prepared as per the standard metallographic practice. Polishing was carried out using emery paper of various grades. Finally, wheel polishing was done with gamma alumina and water as lubricant. The samples were then etched with nital. Relevant microstructures at the various locations of the aluminum weld deposit, weld (Al)-substrate (steel) and the steel substrate were observed under optical microscope.

All metallographic surfaces were also observed under scanning electron microscope (SEM). The SEM micrographs were obtained from secondary electron and back scattered electron images at around 10 kV. Qualitative composition measurements of the elements at the weld substrate interface were conducted by energy dispersive X-ray analysis (EDX) at 10 kV.

3. Results and Discussion

3.1. Interfacial structures

In this section, features of the interfacial structures are shown and discussed on the relation between the welding conditions and the morphology of the interface, formation of the intermetallic compound layer. The interfacial structures are shown in Fig. 3(a), (b) and (c) for ACHF 150 sample and the optical micrographs [Fig. 3(a)] of aluminum weld deposit shows wavy surface at steel substrate. A layer of intermetallic compound and large amount of oxides on weld deposit are also observed. In scanning electron micrographs, a layer of intermetallic compound (Fe_2Al_5) is also clearly observed in Fig. 3(b) and the intermetallic compound of different compositions can be attributed from back scattered electron image [Fig. 3(c)]. Delamination due to poor bonding at steel substrate and aluminum weld deposit are also found in SEM. EDX were carried out at weld deposit, interface and substrate and analysis clearly shows the dilution of Al due to its presence at interface [Fig. 4]. For weld deposits made from DCSP with current settings of 150 Amp, no delamination or poor bonding was observed in optical and scanning electron micrographs for DCSP 150 [Fig. 5] but in DCSP 125 [Fig. 6] some poor bonding at different regions of interface are located.

IOP Conf. Series: Materials Science and Engineering **554** (2019) 012010 doi:10.1088/1757-899X/554/1/012010

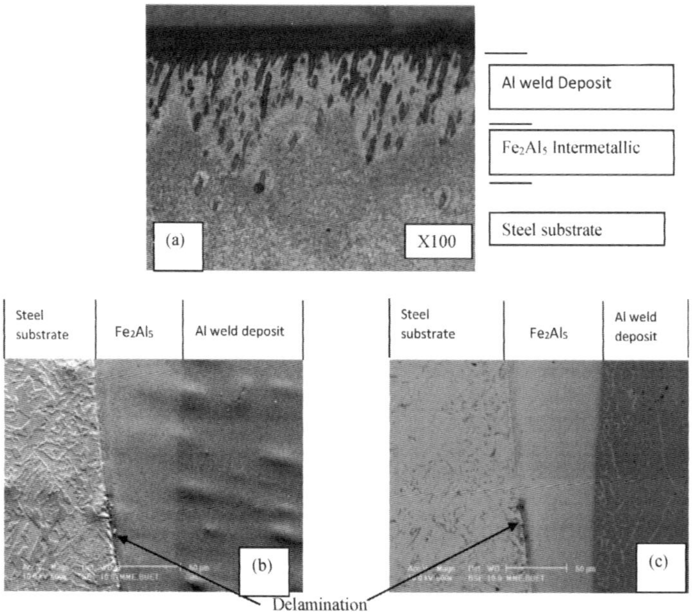

Figure 3. Optical micrograph of ACHF 150 at Al-Steel interface (a), scanning electron micrographs of same region for secondary electron (b) and back scatter electron detection mode (c).

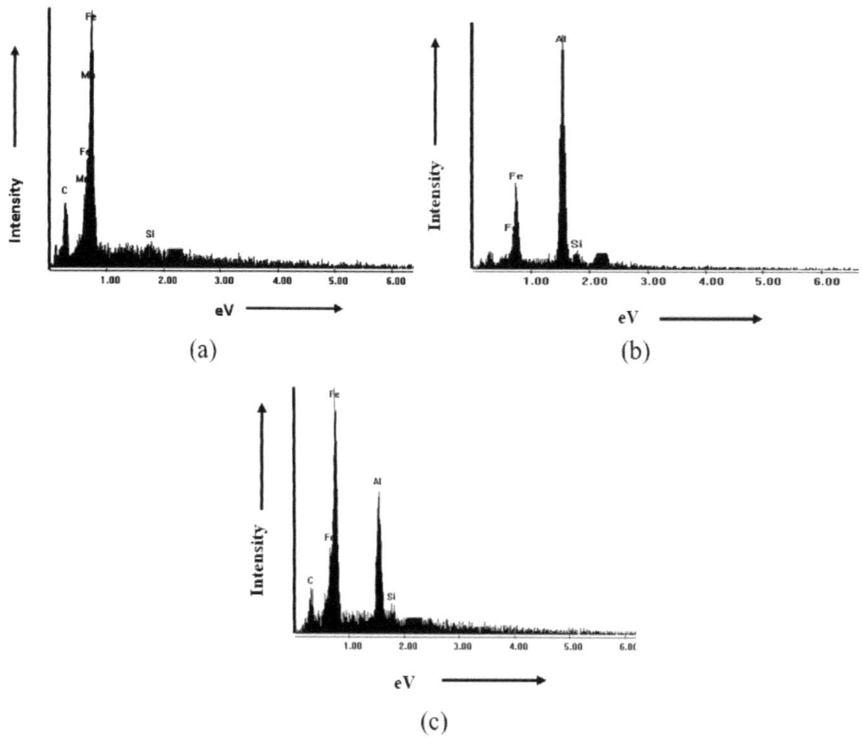

Figure 4. EDX spectra of (a) Steel substrate (b) Al weld deposit and (c) Al/Steel Interface of ACHF 150 sample.

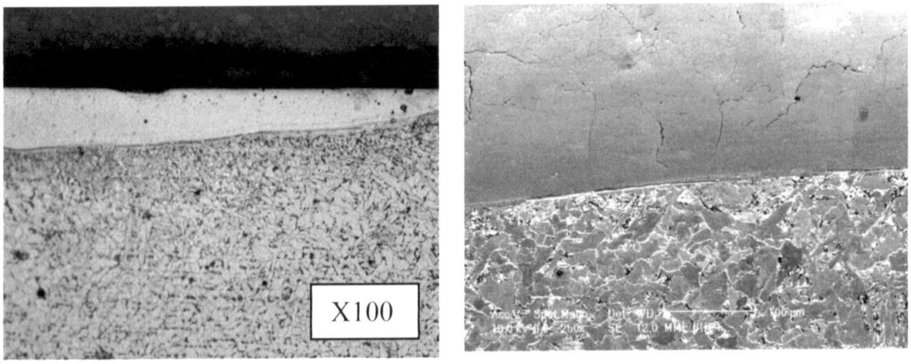

(a) (b)

Figure 5. (a) Optical micrograph of Al-Steel interface of DCSP 150 sample and (b) scanning electron micrographs of same region.

(a) (b)

Figure 6. (a) Optical micrograph of Al-Steel interface of DCSP 125 sample and (b) scanning electron micrographs of same region.

3.2. Relation between ultrasonic wave intensity and irregularity of interface

In this part, the results of the ultrasonic measurement are shown, and the relation between the ultrasonic wave reflected from the surface, the interface and the bottom and the features of the interface are discussed. In immersion method ultrasonic scanning were carried out from both the steel side and aluminum side for three different weld conditions. In order to determine the reflection nature from interfaces the acoustic impedances and the calculated reflected wave ratios at different interfaces are given in Table 3.

Table 3. Acoustic impedance and reflected wave ratios at different interfaces.

Acoustic Impedance, kg/m^2-s			Reflected wave ratio at interface of		
Water (W)	Steel (S)	Aluminum (Al)	W/S	S/Al	Al/W
1.50×10^6	4.54×10^7	1.71×10^7	0.936	-0.453	-0.839

In case of weld deposit made from ACHF 150, the ultrasonic scanning reflection patterns from steel and aluminum sides are shown in Fig 7 (a) and (b) respectively. From steel side the first peak echo at water/steel (W/S) interface is negative whereas, the peak echoes from steel/Al (S/A) and Al/water (A/W) interfaces are positive as the reflected wave ratios for W/S is positive and for S/A and A/W are negative

(Table 3). When the same sample is scanned from aluminum side amplitude of reflected echo from aluminum/steel (A/S) interface [Fig. 7(b)] is almost half of the echo obtained at steel/aluminum interface [Fig. 7(a)] although the reflected echo is almost 50% of the incident echo. This can be explained from the micrographs obtained from optical and scanning electron micrographs (Fig. 3). In optical micrographs [Fig. 3(a)] aluminum weld deposit makes wavy surface at steel substrate. A layer of intermetallic compound and large amount of oxides on weld deposit are also observed. Usually it is reported [18] when a smooth layer of intermetallic was formed, echo reflectivity was in good agreement with the theoretical value whereas, the non-uniform layer caused rough interface due to presence of oxides and the ultrasonic wave was largely scattered at the interface with echo reflectivity decreased below the theoretical value. At present echo reflectivity showed large scattering for ACHF 150 sample and this may due to large amount of oxide may cause poor bonding at steel aluminum interface. In scanning electron micrographs, a thick layer of intermetallic compound (Fe2Al5) was also clearly observed [Fig. 3(b)] and the intermetallic compound of different compositions can be attributed from back scattered electron image [Fig. 3(c)]. Delamination due to poor bonding at steel substrate and aluminum weld deposit are also found in SEM. EDX were carried out at weld deposit, interface and substrate and analysis clearly shows the dilution of Al due to its presence at interface [Fig. 4].

During ultrasonic scanning from steel side, these delamination may cause big reflection [Fig. 7 (a)] due to presence of air between the steel and aluminum as the reflectivity at steel/air is almost -1 due to acoustic impedance of the air is very small. Whereas, when the ultrasonic scanning was carried out from Al side reflections may occur from Al/Fe2Al5 and from delamination at close proximity. As a result, instead of sharp peaks, reflections peaks became broader with low amplitude.

For weld deposits made from DCSP with current settings of 150 amps, the scanning reflection patterns from steel and aluminum sides for same sample are shown in Fig 8(a) and (b) respectively. In general, reflected echoes for DCSP 150 from S/Al interface are of smaller height as compared to ACHF 150 but in some scanning points of DCSP 150 showed similar echo pattern as of ACHF 150. No delamination or poor bonding was observed in optical and scanning electron micrographs for DCSP 150 [Fig. 5] but in DCSP 125 [Fig. 6] some debonding at different regions of interface were located.

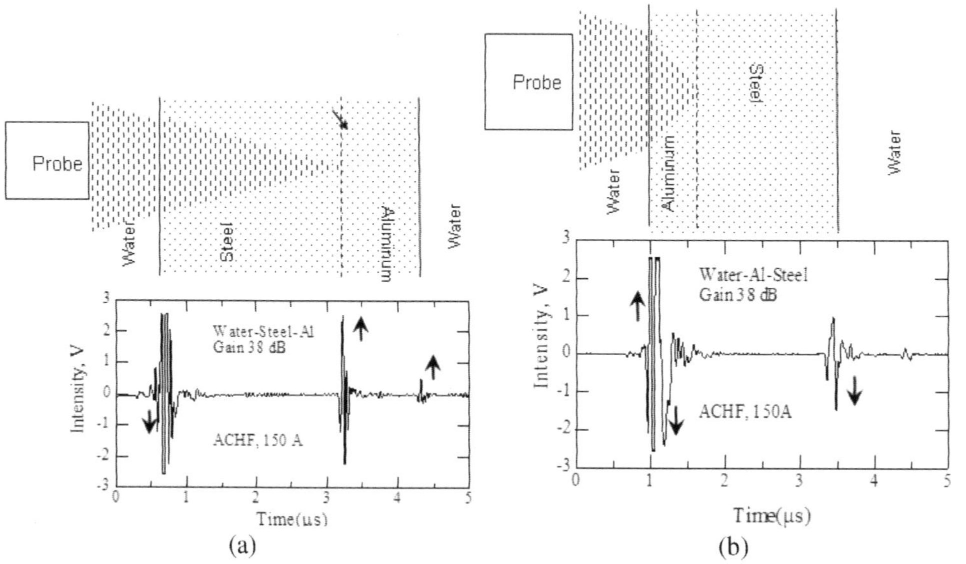

Figure 7. Ultrasonic wave reflection patterns of ACHF 150. Scanning from (a) steel side and (b) aluminum side in immersion method.

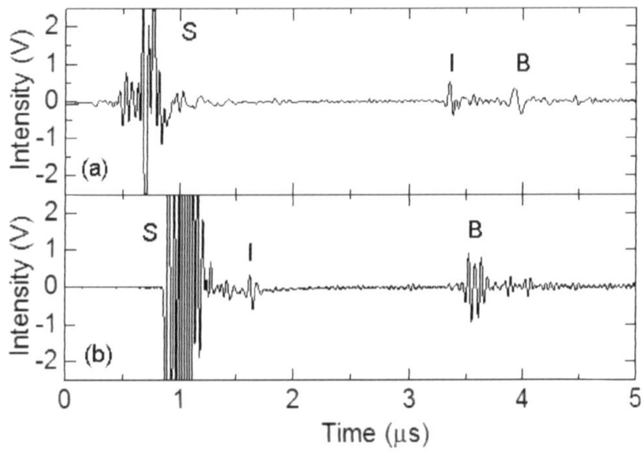

Figure 8. Ultrasonic wave reflection patterns of DCSP 150. Scanning from (a) steel side and (b) aluminum side in immersion method. (S-surface, I-Interface and B-Back surface

3.3. Mapping of the intensity of echo reflected from welding interface

Maps of ultrasonic intensity reflected from the interface obtained by the ultrasonic measurement and also by the scanning acoustic microscope are shown to discuss the relation between the intensity distribution and the bonding conditions. In order to assess the total scanned interfaces of three samples, reflection values at interfaces versus scanning points are plotted and these are shown in Fig. 9. It is clear from these figures that the highest reflection intensities were obtained in ACHF 150, moderate reflection intensities in DCSP 125 and the minimum reflection intensities in DCSP 150.

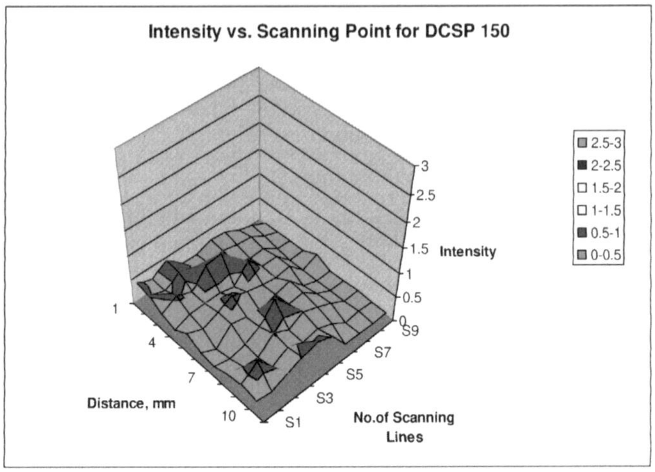

(a)

IOP Conf. Series: Materials Science and Engineering 554 (2019) 012010 doi:10.1088/1757-899X/554/1/012010

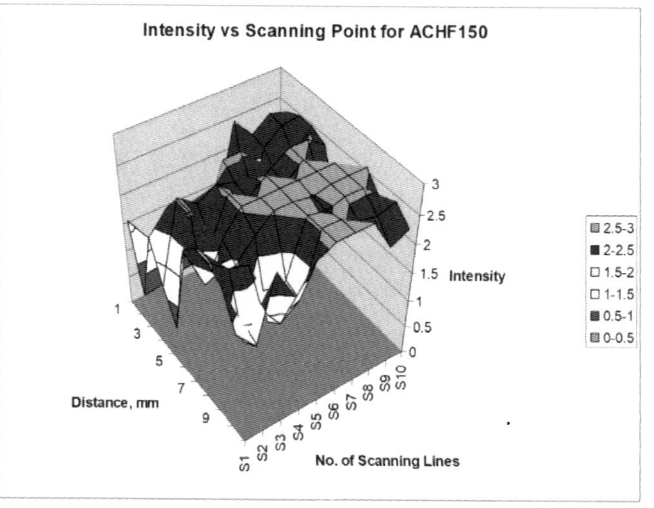

(b)

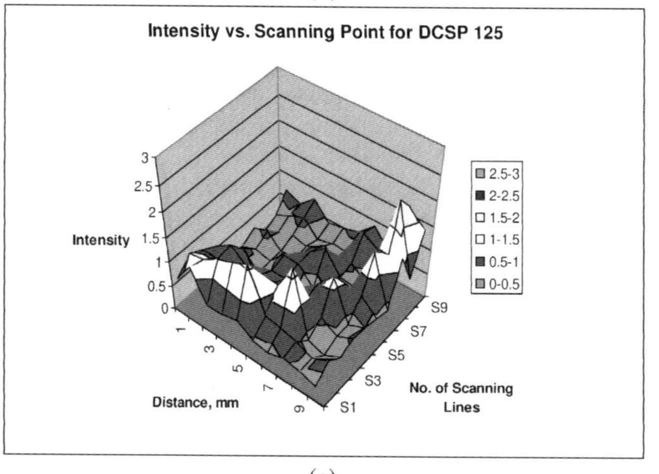

(c)

Figure 9. Mapping of reflected ultrasound echo intensities from (a) ACHF 150, (b) DCSP 150 and (c) DCSP 125 samples in C-scan immersion method.

Finally, bonding conditions of aluminum weld deposits on steel substrates were characterized by scanning acoustic microscope (SAM) with a probe of 50 MHz in order to locate the debonding areas by taking images of aluminum-steel interfaces of three samples. The large bright regions (white spots) in acoustic image were the debonding area as shown in Fig 10. The map of reflected waves gives an idea of how impedance changes within sample. This wave travels to the specimen through a medium or couplant, which is usually deionized water since sound waves could not travel through air at the frequencies used. The ultrasound wave energy traveled through the aluminum at the aluminum's velocity, with a portion of ultrasound energy being reflected back every time when it hit at the aluminum steel interface. ACHF 150 showed the large number of white spots clustered in some regions which are attributed to debonded area [Fig. 10(b)], whereas DCSP 125 revealed white spots in terms of reflection of sound rather scattered in a large area [Fig. 10(f)]. Scanning in DCHP 150 sample almost provided no indication of white spots [Fig. 10(d)] i.e. most of the ultrasound transmitted through the interface which matched with the scanned result of the immersion method as shown in Fig.9(b). As compared to interface images [Fig. 10(b), (d) and (f)], SAM images of the surfaces [Fig. 10(a), (c) and (e)] of three samples only revealed the surface profile of aluminum.

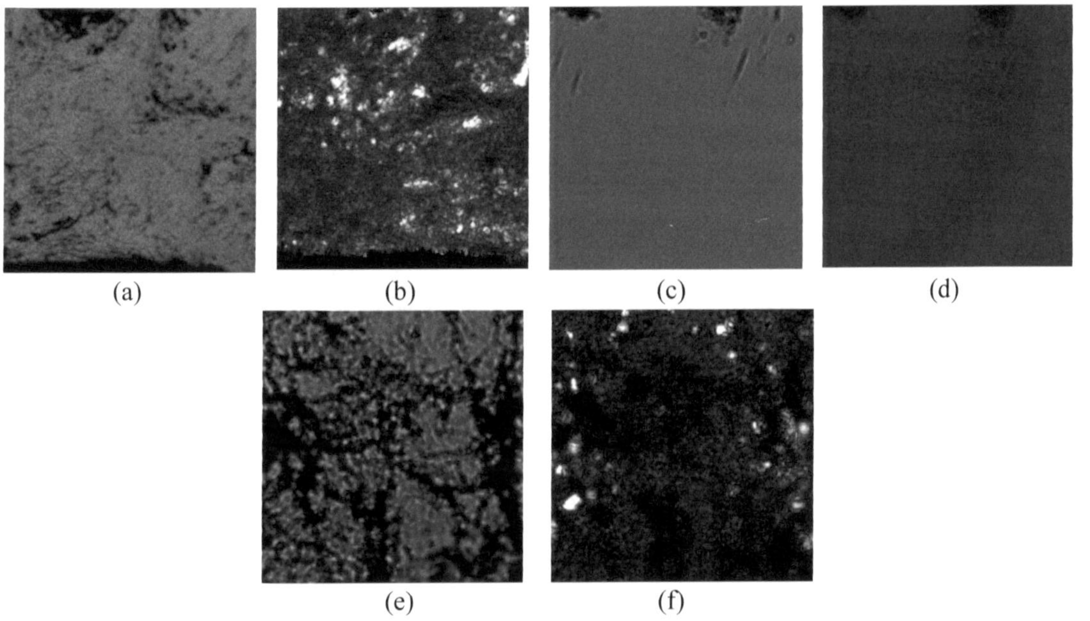

Figure 10. Scanning acoustic microscope images (a) from top and (b) at interface of ACHF 150, (c) from top and (d) at interface of DCSP 150 and (e) from top and (f) at interface of DCSP 125.

4. Conclusions

In this paper, deposition and thus bonding of Al filler on steel substrate by GTAW process at different current settings were evaluated by ultrasound wave propagation studies. From this investigation, the following important conclusion has been derived:

Of the three welded bonds, the joint fabricated by ACHF 150 exhibited intrusion of oxides in bond layer and resulted wavy at Al-Steel interface. Waviness at interface caused weak bonding. Both the ultrasonic immersion-C scan and scanning acoustic microscopy show the maximum reflection in terms of echo height or white spot. Whereas for DCSP 150 where the heat input is maximum associated with stable arc causing good and smooth interface. During ultrasonic inspection DCSP 150 exhibited the minimum reflection resulted from strong bond at Al-Steel interface. When heat input is lower than the DCSP 150 i.e. for DCSP 125 bond at interface was rather poor but superior than the ACHF 150 where arc instability occurred duet to change of polarity during welding. Mapping of ultrasonic C-scanning and image from scanning acoustic microscopy can be used as both qualitative and quantitative tools for characterizing dissimilar metal weld.

Acknowledgement

We greatly acknowledge Universiti Teknikal Malaysia (UTeM) Melaka, Malaysia, Saitama University, Japan and Bangladesh University of Engineering and Technology (BUET) for sponsoring this work. One of the authors is grateful to the Division of Mechanical Science and Engineering, Saitama University for extending the facilities of laboratories and express his sincere thanks to Saitama University for the financial support to carry out this investigation under the cooperative research program.

References

[1] Maruzen. 1990. Welding Handbook. Edited by Japan Welding Society: pp. 496.
[2] Aritoshi, M., and Okita, K. 2002. Friction welding of dissimilar metals. J. of Japan Welding Society 71-6, pp.432–436.

[3] Watanabe, T., and Yoneda, A., et al. 1999. A study on ultrasonic welding of dissimilar metals,1st and 2nd reports. Quarterly J. of Japan Welding Society 17-5, pp. 223–242.
[4] Kouno. 2002. Manufacture of Al/SUS clad by vacuum rolling joining method and its properties. J. Japan Welding Society 71-6, pp.427–431.
[5] Yan, F., Fang, X. F., Chen, L., Wang, C. W., Zhao, J., Chai, F., and Wang, W. 2018. Microstructure evolution and phase transition at the interface of steel/Al dissimilar alloys during Nd: YAG laser welding. Optics and Laser Technology 108, pp 193–201.
[6] Katayama, S. 2002. Dissimilar materials joining by laser. Welding Technology 50-2, pp 69–73.
[7] Malarvizhi, S., and Balasubramanian, V. 2011. Effect of welding processes on AA2219 aluminium alloy joint properties. Trans. Nonferrous Met. Soc. China 21, pp 962-973.
[8] Thomas, W. Metal. Friction stir welding, international patent application PCT/GB92/02203.
[9] Watanabe, T., Takayama, H., and Kimapong, K. 2003. Joining of steel to aluminum alloy by interface-activated adhesion welding. Materials Science forum, 426-432, pp. 4129–4134.
[10] Yoshikawa, K., and Hirano, T. 2001. Numerically controlled friction stir welding in layered dissimilar metal materials of aluminum and steel. Proceedings of the Third Symposium on Friction Stir Welding, Kobe, Japan, September, pp. 1–11.
[11] Kimapong, K, and Watanabe T. 2004, Friction stir welding of aluminum alloy to steel, Welding Journal, Oct., pp 277 – 282.
[12] Pouranvari, M., and Abbasi, M. 2018. Dissimilar gas tungsten arc weld-brazing of Al/steel using Al-Si filler metal: Microstructure and strengthening mechanisms. Journal of Alloys and Compounds 749, pp 121-127.
[13] Su, Y., Hua, X., and Wu, Y. 2013. Effect of input current modes on intermetallic layer and mechanical property of aluminum–steel lap joint obtained by gas metal arc welding. Materials Science & Engineering A 578, pp 340–345.
[14] Rathod, M. J., and Kutsuna M. 2004. Joining of aluminum alloy 5052 and low-carbon steel by laser roll welding. Welding Journal, Welding Research Supplement, pp 16-26 S.
[15] Bobrov V. A., and Khimchenko N. V. 1971. Ultrasonic inspection of clad metals, UDC 620.179.16:62-419.4, Translated From Khimicheskoe I Neftyanoe Mashinstroenie, Aug, 1971, No. 8, pp 29-31.
[16] Hudgell, R. J. 1994. Handbook on the ultrasonic examination of austenitic clad components, The International Institute of Welding and Joint Research Centre, European Commission, Luxembourg.
[17] Kou, S. 2003. Welding Metallurgy, second ed., John Wiley & Sons, Inc., Hoboken, New Jersey.
[18] Kato, H., Abe, S., and Tomizawa T. 1997. Interfacial structures and mechanical properties of steel–Ni and steel–Ti diffusion bonds. Journal of Materials Science 32, pp 5225 – 5232.

Institute of Physics
Dirac House, Temple Back
Bristol BS1 6BE UK

ISSN: 1757-8981
ISBN 978-1-5108-8814-2

71st Northeastern Weed Science Society Annual Meeting 2017

Proceedings – Volume 71

Philadelphia, Pennsylvania, USA
3 – 6 January 2017

NORTHEASTERN
WEED SCIENCE SOCIETY